VEDIC MATHEMATICS

Indispensable to Students Appearing for Competitive Examinations

AF522423

Rajesh Kumar Thakur

B.Sc (Maths Hons.), M.Sc (Maths)
M.Sc (Operation Reseaerch), MA (Education)
ETE, B.Ed

UNICORN BOOKS

Publisher

F-2/16, Ansari Road, Daryaganj, New Delhi-110002
☎ 011-23275434, 23262683, 23250704 • *Fax:* 011-23257790
E-mail: info@unicornbooks.in • *Website:* www.unicornbooks.in

Branch : Mumbai
23-25, Zaoba Wadi, Thakurdwar, Mumbai-401002
☎ 022-22010941, 022-22053387
E-mail: rapidex@bom5.vsnl.net.in

© Copyright: Author

ISBN 978-81-7806-177-1
Vedic Mathematics

Edition: 2019

The Copyright of this book rests with UNICORN BOOKS. No part of this publication may be reproduced, stored in a retrieval system or transmitted, in any form or by any means, electronic, mechanical, photocopying, recording or otherwise, without the prior permission in writing from the Publisher. Anybody doing so shall face legal action and will be liable for damages.

Printed at : Param Offsetters, Okhla, New Delhi-110020

TABLE OF CONTENTS

PREFACE

Mathematics is called the queen of all subjects and the Vedas are the ultimate source of knowledge, the combination of the two is really an astonishing experience for the world. The magical spell of sixteen Vedic sutras, which have brought a revolutionary change in mathematical calculations, cannot be described in a few words.

Mathematics is considered an insipid subject and is disliked by many children, due to its nature of long calculations. This book on Vedic Mathematics brings the students in a comfortable zone as far as mathematical calculation is concerned. The ability to perform basic mathematical calculations mentally is an art and this book serves this very purpose.

Every topic written in this book is self-explanatory. For the benefit of non-maths students, the book takes care to explain oft-used terms in simple language.

I would extend my thanks to all those who have played a role in making this available to the reader. I specially thank the staff of Unicorn Books for extending their constant support and encouragement.

In spite of all efforts, some errors might have crept in. I shall be grateful to all my readers if the same are brought to my notice. Suggestions for improvement of the book will be gratefully acknowledged.

—Rajesh Kumar Thakur
rkthakur1974@gmail.com

Introduction to Vedic Mathematics

यथा शिखा मयूराणां नागानां मणयो यथा ।
तद्वद्वेड्ग-शास्त्रणां, गणितं मूर्धानि स्थितम् ॥

(Like the crest on the heads of peacocks, like the gems on the hoods of cobras, mathematics is at the top of Vedanga Jyotisha)

Vedic Mathematics may sound a new name for some of you but it was introduced around 50 years ago by Late Sankaracarya Bharti Krishna Tirthaji Maharaj.

The word 'Veda' literally means the 'fountain head and illimitable storehouse of all knowledge'. Therefore, Vedas contain all knowledge which is essential for mankind. The original book of Vedic Mathematics written by His Holiness Swami Bharti Krishna Tirthaji Maharaj has 16 sutras and 13 sub-sutras all together. The original source of Vedic Mathematics is the Atharva Veda and all the sutras and sub-sutras were discovered by Sri Swamiji himself.

This present book is a small effort to bring a revolutionary change in the minds of readers who find Mathematics an insipid subject. The main reason behind Mathematics phobia is lengthy calculation. The traditional method of calculation is not user-friendly and there is hardly any room for choice and experimentation. Addition, Subtraction, Multiplication, Division, Square, Square Root, Cube, Cube Root are the basic structures of mathematics.

Vedic Mathematics presents varieties of methods which one can use according to one's need. Not only that, this book has two more feathers in its cap: one is **Casting out Nines,** a unique and special method to check the accuracy of Addition, Subtraction, Multiplication and Division in a few seconds and the second special feature is the topic on solving **Simple Equations.** Though *Vedic Mathematics* has many things to offer but keeping in view the basic needs of the present-day student who wants faster calculation technique with 100% accuracy, this book has been designed accordingly.

Vedic Mathematics opens a new horizon for the mathematics lover. It presents the wider platform to experiment with. It is unique in many ways:

- It is 5-6 times faster than the traditional method.
- It helps to carry out tedious and cumbersome arithmetical operations mentally.
- This is truly unconventional and lucid.
- It saves our time and precious energy.
- The accuracy of the problem can be judged in seconds.

These unique methods of calculation are easy to understand; and the best feature of *Vedic Mathematics* is its simple presentation. Even the common man who is not much interested in Mathematics and suffers from Mathematical phobia can find the presentation of *Vedic Mathematics* interesting.

❑❑❑

VEDIC SUTRAS

- Ekadhikena Purvena एकाधिकेन पूर्वेण
- Nikhilam Navatascaramam Dasatah निखिलम् नवतश्चरमं दशत:
- Urdhva Tiryagbhyam उर्ध्वतिर्यभ्याम्
- Paravartya Yojayet परावर्तयोजयेत
- Sunyam Samya Samuccaye शून्यं साम्य समुच्चये
- Anurupye Sunyamanyat आनुरूप्ये शून्यमन्यत्
- Sankalana-Vyavakalanabhyam संकलन-व्यवकलनाभ्याम्
- Puranapuranabhyam पूरणापूरणाभ्याम्
- Calanakalanabhyam चलनकलनाभ्याम्
- Yavadunam यावदूनम्
- Vyastisamastih व्यष्टि समष्टि
- Sesanyankena Caramena शेषाण्यंकेन चरमेण
- Sopantya Dvayamantyam सोपान्त्य द्वयमन्त्यम्
- Ekanyunena Purvena एकन्यूनेन पूर्वेण
- Gunita Samuccayah गुणित समुच्चय:
- Gunaka Samuccayah गुणक समुच्चय:

VEDIC SUB-SUTRAS

Anurupyena आनुरूप्येण

Sisyate Sesasamjnah शिष्यते शेषसंज्ञ

Adyamadyenantya-Mantyena आद्यमाद्येनान्त्यमन्त्येन

Kevalaih Saptakam Gunyat केवलैः सप्तकं गुण्यात

Vestanam वेष्टनम्

Yavadunam-Tavadunam यावदूनम तावदूनम

Yavadunam Tavadunikrtya Varganaca Yojayet यावदूनम तावदूनम कृत्य वर्ग च योजयेत्

Antyayotdasakepi अन्त्ययोर्दशकेऽपि

Antyayoreva अन्त्ययोरेव

Samuccaya Gunitah सामुच्चय गुणितः

Lopanasthapanabhyam लोपस्थापनाभ्याम्

Vilokanam विलोकनम्

Gunita Samuccayah Samuccaya Gunitah गुणित समुच्चयः समुच्चयगुणितः

1

ADDITION

Introduction

In addition, we take only two factors under consideration:

(i) speed

(ii) accuracy

The conventional method taught in our school for a long time, though suitable lacks speed. There is a greater chance of mistake. The main problem one faces in addition is transferring the carry number from one column to another. To avoid this absurdity and to experiment in a unique way, Vedic method discussed here has a higher degree of accuracy and is much faster than the traditional method.

Vedic Sutras for Addition:

This book deals with only two Vedic sutras for addition.

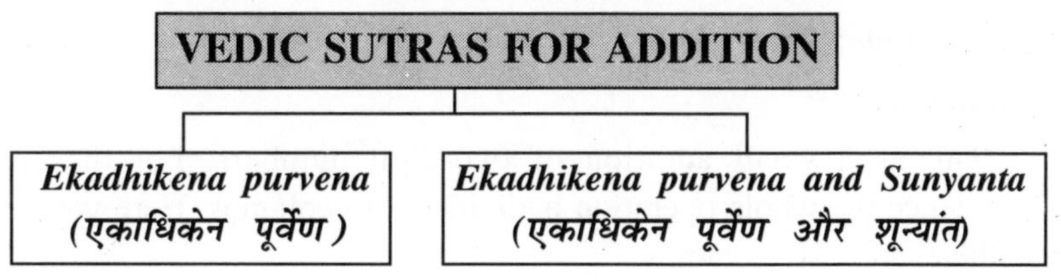

Meaning of Vedic Sutras

1. Ekadhikena Purvena *(एकाधिकेन पूर्वेण)*: It means **"one more than the previous one."**

   Example: Ekadhika of 13 = 13+1 = 14

2. Sunyanta (*शून्यांत*): It means **"ending with zero"**

   Example: 98 = 100 – 2 or 90 + 8

   Here 98 is partitioned in two ways each having zero at the end.

Addition by Vedic Methods

There are several methods under Vedic Mathematics for addition but the best two out of them have been discussed above. Here we shall take a few examples under each subtitle and discuss the procedure under it.

Ekadhikena purvena *(एकाधिकेन पूर्वेण)* method:

The whole procedure of adding can be summarised in the following steps/rules:

(a) Add the unit digits column wise.

(b) When the running total becomes greater than 10 put a dot or tick to that number.

(c) Move ahead with the excess of ten and add it to the next digit of the column.

(d) Lastly count the number of ticks and place it down to the number next to the unit place figure and sum the two.

Let us take a few examples:

Example: 96 + 65 = ?

Explanation:

This is a small addition of two-digit numbers. For the sake of convenience the whole operation is divided into columns. Here you can see that there are two columns.

2nd column	1st column
9	6
6	5

Let us start adding the digits of 1st column downwards. Always mind one thing that when the running total becomes 10 or greater put a dot and move ahead with the excess of ten and add it to the next digit of the column.

In the first column: 6 + 5 = 11, so excess of ten (11 – 10 = 1) is written in the remainder column, putting a dot on 5 in the first column.

In the second column: 9 + 6 = 15, so excess of ten (15 – 10 = 5) is written in the remainder column putting a dot on 6.

Once the running total of all the columns is finished, count the number of dot(s) of each column and place it down to the number next to the unit place figure and **Add** to get the final sum.

Solution:

		9	6
		•	•
		+ 6	5
Running total		5	1
Number of dots	1	1	
	1	6	1

Example: 326 + 654 + 987 = ?

Solution:

	3	2	6
			•
	6	5	4
	•	•	
	9	8	7
Running total	8	5	7

Explanation:

In the first column: 6 + 4 = 10, so take away the excess 0 (10 – 10 = 0) and add it with the next digit 7 of the first column. i.e. 0 + 7 = 7. Write the final sum 7 in the remainder column. Put a dot over 4.

In the second column: 2 + 5 + 8 = 15, so write the excess 5 (15 – 10 = 5) in the remainder column. Put a dot over 8.

In the third column: 3 + 6 + 9 = 18, so write the excess 8(18 – 10 = 8) in the remainder column. Put a dot over 9.

Now count the number of dots in each column and place it down to the number next to the unit place as done in the first example and sum the two to get the final result.

Running total		8	5	7
Dots	1	1	1	
Sum	1	9	6	7

Example: 6489 + 5642 + 3241 = ?

Solution:

6	4	8	9
·	·	·	·
5	6	4	2
+ 3	2	4	1
4	2	6	2

Explanation:

In the first column: 9 + 2 = 11, so place a dot over 2 and add the excess 1 (11 – 10 = 1) to the next figure of the first column, i.e. 1 + 1 = 2

In the second column: 8 + 4 = 12, so place a dot over 4 and add the excess 2 (12 – 10 = 2) to the next digit of the second column, i.e. 2 + 4 = 6

In the third column: 4 + 6 = 10, so place a dot over 6 and add the excess 0 to the next digit of the column, i.e. 0 + 2 = 2

In the fourth column: 6 + 5 = 11, so place a dot over 5 and add the excess 1(11 – 10 = 1) to the next digit of the column, i.e. 1 + 3 = 4

Now count the number of dots in each column and place it down to the number next to the unit place as done in the first example and sum the two to get the final result.

Running total		4	2	6	2
Dots	1	1	1	1	
Sum	1	5	3	7	2

Having discussed the examples of 2, 3 and 4 digits in detail; let us move to the bigger examples. You have so far noticed that the excess of ten which has been frequently used here is nothing but the carry of the unit digit of the sum total. Suppose we add 4 + 8 = 12, so the excess of ten is nothing but the unit digit 2 because 12 – 10 = 2.

Example: 87643 + 84397 + 38549 + 29765 = ?

Solution:

	8	7	6	4	3
	•	•		•	•
	8	4	3	9	7
			•		
	3	8	5	4	9
	•	•	•	•	•
	+ 2	9	7	6	5
	1	8	1	3	4
Dots	2	2	2	2	2

Adding the running total and dots in the following manner as discussed in rule (d):

Running total		1	8	1	3	4
Dots	2	2	2	2	2	
Total	2	4	0	3	5	4

Explanation:

In the first column:

- 3 + 7 = 10 so a dot is put over 7 and the excess (10 – 10 = 0) is carried to the next number in the same column.
- 0 + 9 + 5 = 14 so a dot is placed over 5 and the excess 4(14 – 10 = 4) is written below the line in the running total.

In the second column:

- 4 + 9 = 13 so a dot is placed over 9 and the excess 3(13 – 10 = 3) is carried to the next number.
- 3 + 4 + 6 = 13 so the next dot is placed over 6 and excess 3 (13 – 10 = 3) is written below the line.

In the third column:

- 6 + 3 + 5 = 14 so the dot is placed over 5 and excess 4, i.e. (14 – 10 = 4) is added to the next digit 7.
- 4 + 7 = 11 so the next dot is placed over 7 and excess 1, i.e. (11 – 10 = 1) is written below the line.

In the fourth column:

- 7 + 4 = 11 place the dot over 4 and add the excess 1.
- 1 + 8 + 9 = 18 so place the dot over 9 and write the excess 8 below the line.

In the fifth column:

- 8 + 8 = 16. Place the dot over 8 and add the excess to the next number of the same column.
- 6 + 3 + 2 = 11, place the dot over 2 and excess 1 (11 – 10 = 1) is finally placed as the running total.

Once the addition of the entire column is finished we have to count the number of dots in the respective columns and place it down to the number next to unit place figure. The sum of the two gives us the grand total.

Example: 3 7 9 9 5 + 6 8 9 8 6 + 7 5 4 3 8 + 5 8 9 0 9 = ?

Solution:

	3	7	9	9	5
		•	•	•	•
	6	8	9	8	6
	•	•	•	•	
	7	5	4	3	8
	•		•		•
+	5	8	9	0	9
Running Total =	1	8	1	0	8
Dots =	2	2	3	2	2

Arranging the running total and dots according to rule (d)

		1	8	1	0	8
+	2	2	3	2	2	
	2	4	1	3	2	8

Explanation:

In the first column:

- 5 + 6 = 11 so put a dot over 6 and move ahead with the excess 1.
- 1 + 8 + 9 = 18 so write 8 as the running total and move to the next column.
- Proceed the same way until the running total of all the columns is done.
- Count dots of each column and place it below the running total according to rule (d) to get the final answer.

Improvised Ekadhikena Method

In order to save some more time you can put the dot to the number next to the digit where the running total is greater than 10. This tick will enrich that number by 1.

i.e. $\dot{9}$ = 9 + 1 = 10 (Ekadhikena Purvena)

This will save some extra time of counting the dots and again adding it to the running total. Let us understand the entire process with the following examples.

Example:

$$
\begin{array}{ccccccc}
 & 3 & 2 & 9 & 7 & 3 & 6 \\
 & & \bullet & \bullet & & \bullet & \\
 & 4 & 6 & 5 & 7 & 2 & 8 \\
\bullet & \bullet & & \bullet & \bullet & \bullet & \\
0 & 6 & 2 & 3 & 9 & 9 & 9 \\
 & & \bullet & & & & \\
+ & 5 & 5 & 4 & 3 & 2 & 1 \downarrow \\
\hline
1 & 9 & 7 & 3 & 7 & 8 & 4
\end{array}
$$

Explanation: Addition for each column will be done as you generally do in the normal addition. For the convenience of readers, the arrow has been shown moving downwards.

In the first column:

- 6 + 8 = 14 > 10 so place a dot over 2 ($\dot{2}$), number preceding 8. Take the excess 4 (14 – 10 = 4) to the next number of 1st column.
- 4 + 9 = 13 >10 so place a dot over 9, number preceding 9 of the first column and take the excess 3 (13 – 10 = 3) to the next number of the 1st column.
- 3 + 1 = 4 since there is no more digits left in the first column so final sum 4 should be written at the bottom line.

In the second column:

- 3 + $\dot{2}$ (=3) + $\dot{9}$ (=10) = 16 > 10 so place a dot over 9 in the third column and take the excess 6(16 – 10 = 6) to the next number of the second column.
- 6 + 2 = 8. Since there is no more digits left in the second column so the final sum of the second column is written below the line.

The sum of the rest of the columns will be done accordingly.

Suppose you encounter a case when the sum of the digits is more than 10 and there is no column left, what will you do?

Very simple, put a zero behind the number so that dot of Ekadhikena can be placed over it.

In the sixth column 3 + 4 + 6 = 14 >10, so a dot needs to be put to the preceding number in the seventh column hence a zero (**0**) is placed in the seventh column.

Let us take another example with some explanatory notes to understand the whole process of addition in improvised Ekadhikena method.

Example: 124 + 4234 + 8238 + 646 + 5321 + 3510 + 8989 = ?

Solution:

$$
\begin{array}{rccccc}
 & & & 1 & 2 & 4 \\
 & & 4 & 2 & 3 & 4 \\
 & \dot{0} & 8 & 2 & \dot{3} & 8 \\
 & & \dot{0} & \dot{6} & \dot{4} & 6 \\
 & & 5 & 3 & 2 & 1 \\
 & \dot{0} & \dot{3} & 5 & 1 & 0 \\
+ & \dot{0} & \dot{8} & \dot{9} & \dot{8} & 9 \\
\hline
 & \mathbf{3} & \mathbf{1} & \mathbf{0} & \mathbf{6} & \mathbf{2}
\end{array}
$$

Explanation:

In the first column:

- 4 + 4 + 8 = 16. It is more than 10, so a dot is placed next to 8, i.e. on 3. Here dot on 3 makes it one richer by itself. 3 = 3 +1 = 4

- Take the excess 6 (16 – 10 = 6) and move to the next digit. The sum of 6 + 6 = 12 >10. So place a dot on 4, next to 6 in the preceding column.
- Again 2(excess of the previous stage) + 1 + 0 + 9 = 12, so placing a dot on 8 write the final 2 at the end of first column.

In the second column:

- $2 + 3 + \dot{3}(= 4) + \dot{4}(= 5) = 14$ is more than 10, so dot is placed to the preceding number 6 in the third column.
- Proceed with the excess 14 – 10 = 4 and add the remaining digits in the second column.

$4 + 2 + 1 + \dot{8}\ (= 9) = 16$

- Write the excess 6 (16 – 10) below the second column and move to the third column likewise.

This process will be continued till every column is taken into account. Once you practice the examples of the first method, it will be easier for you to understand the Improvised method.

Ekadhikena Purvena and Sunyanta (एकाधिकेन पूर्वेण और शून्यांत)

This method is the combination of two Vedic sutras – Ekadhikena Purvena and Sunyanta **(एकाधिकेन पूर्वेण और शून्यांत).** Once you are very much comfortable with the previous two methods, you can take a great advantage of this club method. This is faster than the two discussed in the beginning. Before we take an example to illustrate this method, let us discuss how this method works.

Procedure:

1. Take at least two-digit numbers in each intersection of rows and columns.
2. While summing up the individual column, bring one number as the multiple of 10.

3. The deficit or excess should be subtracted or added to the next number of that column.
4. When the sum is greater than 100, place a dot on the digit next to it in the preceding column as done in Improvised Method.
5. Follow the same procedure for the remaining column.

Example: Add 254 + 368 + 596 = ?

Solution: Place the number in the column as shown here

$$\begin{array}{rcc} & \mathbf{2^{nd}} & \mathbf{1^{st}} \\ & 2 & 54 \\ & \dot{3} & 68 \\ + & \dot{5} & 96 \\ \hline & 12 & 18 \end{array}$$

Explanation:

In the first column:

(a) 54 + 68 = 50 + 4 + 60 + 8

= (50 + 60) + 4 + 8

= 110 + 12 = 122

(b) Since 122 > 100, so place a dot over 3

(c) Add the excess 22 (122 – 100 = 22) to the next number 96.

22 + 96 = 20 + 2 + 90+ 6

= (20 + 90) + 2+ 6

= 110 + 8 = 118

(d) Since 118 > 100, so place a dot over 5 (**Rule 4**) and write the excess. 18 (118 – 100) in the answer column below the line.

In the second column:

$$2 + \dot{3} + \dot{5} = 2 + 4 + 6 = 12$$

Since there is no more digits left so the final 12 will be placed in the answer column.

Example: Add 7534 + 2459 + 1932 + 6547 = ?

Solution: Arranging the numbers in the column, taking two digits in each column.

$$
\begin{array}{rrrr}
 & & 2^{nd} & 1^{st} \\
 & & 75 & 34 \\
 & & 24 & 59 \\
 & \dot{0} & \dot{19} & 32 \\
+ & & 65 & 47 \\
\hline
 & 1 & 84 & 72
\end{array}
$$

Explanation:

Column 1:

$$
\begin{aligned}
34 + 59 + 32 &= 30 + 4 + 50 + 9 + 30 + 2 \\
&= (30 + 50 + 30) + 4 + 9 + 2 \\
&= 110 + 10 + 5 = 120 + 5 = 125
\end{aligned}
$$

Since 125 > 100, so place a dot over 9 in the preceding column and move with the excess (125 – 100 = 25) to the next number to be added.

$$25 + 47 = 72$$

Column 2:

$$
\begin{aligned}
& 75 + 24 + \dot{19}\ (= 20) \\
&= 70 + 5 + 20 + 4 + 20 \\
&= 70 + 20 + 20 + 9 = 119
\end{aligned}
$$

Since 119 >100, so place a dot to the preceding number in the third column. Since there is no digit left, put a zero (0) arbitary in the third column and place a dot over it.

Add excess 19 (119 – 100 = 19) with 65.

$$19 + 65 = (20 - 1) + (60 + 5) = 84$$

Place this 84 below the line in the second column.

Column 3: $\dot{0} = 1$

Write 1 in the third column. Hence the final sum is 18472.

Example: 112065 + 360051 + 43782 + 156450 = ?

Solution:

Divide the number in columns, taking two digits at a time.

3rd	2nd	1st
11	20	65
36	$\dot{00}$	51
04	37	82
$\dot{15}$	$\dot{64}$	50
67	23	48

Explanation:

Column 1:

65 + 51 = 60 + 5 + 50 + 1
= 60 + 50+ 5 + 1
= 110 + 6 = 116 > 100

- Place a dot over 0 in the column 2. (**Rule 4)**
- Add the excess (116 – 100 = 16) with the next two-digit number 82.
  16 + 82 = 10+ 6+ 80 + 2 = 98
  98 + 50 = 100 – 2 + 50 = 100 + 50 – 2 = 148. Since 148 > 100, so place a dot over 4 in the second column and write 48 in the answer column below the line.

Column 2:

$\dot{20} + 00$ —

$$20 + \dot{00}\ (= 1) + 37 + \dot{64}\ (= 65)$$
$$= 20 + 1 + 30 + 7 + 60 + 5$$
$$= 20 + 30 + 60 + 1 + 7 + 5$$
$$= 110 + 13$$
$$= 123$$

Since 123 >100, so put a dot over 5 in the third column and write 23 in the answer column.

Column 3:

$$11 + 36 + 4 + 1\dot{5} \ (= 16)$$
$$= 10 + 1 + 30 + 6 + 4 + 10 + 6$$
$$= (10 + 30 + 10) + 1 + 6 + 6 + 4$$
$$= 50 + \ 17 = 50 + 10 + 7 = 67$$

Write 67 in the answer column.

The two Vedic sutras discussed above are not limited to simple addition only but can be used in the following type of sums:

- Kilogram - grams
- Rupee - paisa
- Decimal
- Hour - minute

Let us take a few more examples.

Example: Add 112 Kg 65 g, 360 Kg 85 g, 284 Kg 678 g, 154 Kg 469 g and 72 Kg 424 g.

Solution:

- Make two columns; namely Kg and g
- Since 1Kg = 1000g, so place three-digit number under gram column.
- Apply any of the above method.

	Kg	**g**
	112	065
		··
	360	085
	·	·
	284	678
	··	·
	154	469
		··
+	72	424
	983	721

Explanation:

- Here 65 gram and 85 gram are written as 065 and 085.
- Start adding from the unit column
- 5 + 5 = 10; so place a dot over 8

  0 + 8 + 9 = 17 > 10, place a dot over 6

  7 + 4 = 11 >10, place a dot over 2.
- Write 1 to the unit column and proceed in the same fashion to find the sum of all the columns. The sum of all the columns gives the result.

Example: Add Rs 28 40 P, Rs 32 04 P, Rs 65 78 P, Rs 38 92 P and Rs 16 92 P

Solution: We all know Re 1 = 100 Paisa , so 4p and 7p should be written as 04 p and 07 p. Here too we need to make two columns under the heading Rs and P for Rupees and Paisa . The rest of the procedure is the same as the simple addition is done. You can apply any of the three methods discussed so far and enjoy the beauty of Vedic calculation for addition.

	Rs	P
	28	40
	32	04
	• •	
	065	78
	•	
	38	92
	•	
+	16	92
	182	06

In the first column:

(a) 40 + 04 + 78 = 40 + 4 + 70 + 8

= (40 + 70) + 4 + 8

= 110 + 4 + 8

= 122

Since 122>100, take the excess 22(122 – 100) and add it with the next number of the first column. Place a dot as directed on 5 in the preceding column.

(b) $22 + 92 = 20 + 90 + 2 + 2$

$= (10 + 10 + 90) + 2 + 2$

$= 110 + 4 = 114$

Since 114 > 100, so place a dot over 8 in the preceding column and add the excess 14 (114 – 100 = 14) with the next number of the 1st column.

(c) $14 + 92 = 10 + 4 + 90 + 2$

$= (10 + 90) + 4 + 2$

$= 100 + 6 = 106$

106 > 100, so place a dot over 6 in the preceding column and write the excess in the answer column because there is no more digit left in the first column to be added.

In the second column:

$$
\begin{aligned}
28 + 32 + 6\dot{5} &= 20 + 8 + 32 + 60 + 6 \\
&= 20 + (32 + 8) + 60 + 6 \\
&= 20 + 40 + 60 + 6 \\
&= 20 + (40 + 60) + 6 \\
&= 20 + 100 + 6 = 126
\end{aligned}
$$

Since 126 > 100, place a dot over 0 (As there is no pair of digits left in the second column so a zero has been placed before it.) Move with the excess 26 (126 – 100 = 26) and add it with the next number of the second column.

$$
\begin{aligned}
26 + 3\dot{8} + 1\dot{6} &= 20 + 6 + 30 + 9 + 10 + 7 \\
&= (20 + 30 + 10) + 6 + 9 + 7 \\
&= 60 + 1 + 5 + 10 - 1 + 5 + 2 \\
&= 82
\end{aligned}
$$

Write 82 in the answer column.

In the third column:

$\dot{0} = 1$

Final sum = **Rs.**182 06**P**

Try some more examples of different types and practice them until you find yourself equally comfortable with Vedic Addition technique.

EXERCISE FOR PRACTICE

Exercise 1.1

Add the following by Ekadhikena Purvena Sutras:

1.		2.	
	75934		876549762
	87628		345982769
	34879		470154897
	14093		284579657
+	256	+	145469885

3.		4.	
	762874		3476928
	593487		274039
	4789		8752684
+	187049	+	48998

5.	
	6938789159689
	5248792300000
	7895248301554
	8736200145932
+	5260148530489

Exercise 1.2

Add the following by Ekadhikena and Sunyanta Method:

1.
```
   5892458
   2547890
   7891642
   8764501
+    45321
```

2.
```
   6589711256
   9854721036
   5548760025
   9405547155
+     2568790
```

3.
```
   587
   256
   875
   250
   569
+  253
```

4
```
   2568
   9856
   8555
   7426
   7840
+  3654
```

5.
```
   2654793
   6501547
   8725780
   3571598
   1729617
      6540
```

Exercise 1.3

Add the following:

1.

	Km	m
	256	145
	253	874
	326	099
+	2	450

2.

	Kg	g
	87	958
	11	025
	05	652
+	1	006

3.

	Rs	P
	325	25
	625	32
	325	62
+	2	06
	28	65

4.

	Rs	P
	659	38
	968	15
	159	64
+	6	98
	98	36

2

SUBTRACTION

Introduction

The process of finding a quantity which when added to one of the two given quantities will give the other. These quantities are respectively the subtrahend and minuend. The final result is called the remainder.

Example:

4 2 7 8	Minuend
– 3 1 2 1	Subtrahend
1 1 5 7	Remainder

Subtraction is considered a difficult mathematical operation. Vedic method discussed here will help you to ease the very concept of subtraction and even the difficult-looking subtraction can be done in a few seconds mentally.

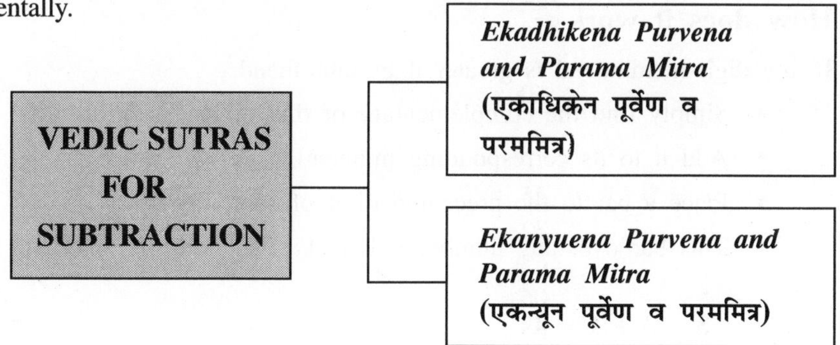

Meaning of Vedic Sutras

Ekadhikena Purvena and Parama Mitra (Complementary) Method (एकाधिकेन पूर्वेण व परममित्र विधि): This method is the combination of two sutras Ekadhikena and Parama Mitra. The first sutra Ekadhikena means one more than the previous one. A detailed discussion is given in Chapter 1. The second sutra Param Mitra literally means the complementary. Two digits are said to be complementary to each other if their sum is 10.

Ekanyuena Purvena and Parama Mitra Method (एकन्यून पूर्वेण व परममित्र विधि): This formula is also the combination of two sutras Ekanyuena Purvena and Param Mitra. The first one literally means one less than the previous one. As far as the complementary method is concerned, it is the sum of the two digits giving the final sum as 10.

0 and 10 are complementary to each other as $0 + 10 = 10$

1 and 9 are complementary to each other as $1 + 9 = 10$.

2 and 8 are complementary to each other as $2 + 8 = 10$.

3 and 7 are complementary to each other as $3 + 7 = 10$.

4 and 6 are complementary to each other as $4 + 6 = 10$.

5 and 5 are complementary to each other as $5 + 5 = 10$.

Why Subtraction by Vedic Method

The best advantage of Vedic method over any other method is that it avoids borrowing or carrying. It simply works like the general addition.

How does it work

If any digit in minuend is greater than subtrahend –

- Simply find the complementary of that digit at subtrahend.
- Add it to its corresponding minuend.
- Place a bar to the preceding digit of subtrahend.
- This bar over any number will make that number richer by 1.

i.e. $\mathbf{\bar{3} = 3 + 1 = 4}$

□□□

Let us take a few examples to understand the Vedic Subtraction clearly.

Example: Subtract 19 from 26

Solution:

$$\begin{array}{cc} 2 & 6 \\ -\overline{1} & 9 \\ \hline \times & 7 \end{array}$$

Explanation:

1. 6 – 9 is not possible with (+) sign, so take the complementary of 9 and add it with 6.

   i.e. **1** (complementary of 9) **+ 6 = 7**

2. Apply Ekadhikena Purvena sutras and place a bar over 1, a digit behind 9 at **the subtrahend,** making it **1** richer than itself.

   i.e. $\overline{1} = 1 + 1 = 2$

3. Now $2 - \overline{1} = 2 - 2 = 0$

   Hence the operation completes.

Example: Subtract 185 from 534

Solution:

$$\begin{array}{cccc} & 5 & 3 & 4 \\ - & \overline{1} & \overline{8} & 5 \\ \hline & 3 & 4 & 9 \end{array}$$

Explanation:

1. 4 – 5 is not possible with (+) sign, hence add the complementary of 5 to 4. Hence 4 – 5 = 4 + 5 (complementary of 5) = 4 + 5 = 9. Place a bar over 8, the preceding digit of 5.

2. $3 - \overline{8} = 3 - 9$ is also not possible with (+) sign, hence use complementary of 9 and add it to 3

   i.e. 3 – 9 = 3 + 1 (complementary of 9) = 4 Place a bar over 1.

3. $5 - \overline{1} = 3$. Since no more digits are left so the operation completes.

Example: Subtract 3782 from 8543

Solution:

$$\begin{array}{r} 8\ \ 5\ \ 4\ \ 3 \\ -\overline{3}\ \ \overline{7}\ \ 8\ \ 2 \\ \hline 4\ \ 7\ \ 6\ \ 1 \end{array}$$

Explanation:

1. 3 – 2 = 1
2. 4 – 8 is not possible with (+) sign, hence 4 – 8 = 4 + complement of 8 = 4 + 2 = 6. Put a bar over 7, the preceding of 8 making it 1 richer by itself.
3. 5 – $\overline{7}$ = 5 – 8 = 5 + complement of 8 = 5 + 2 = 7. Put a bar over the preceding digit 3.
4. 8 – $\overline{3}$ = 8 – 4 = 4

Example: Subtract 49875 from 28999

Solution:

$$\begin{array}{r} 4\ \ 9\ \ 8\ \ 7\ \ 5 \\ -2\ \ \overline{8}\ \ \overline{9}\ \ \overline{9}\ \ 9 \\ \hline 2\ \ 0\ \ 8\ \ 7\ \ 6 \end{array}$$

Explanation:

1. 5 – 9 = 5 + 1 (complementary of 9) = 6, put a bar on preceding 9.
2. 7 – $\overline{9}$ = 7 – 10 = 7 + 0 (complementary of 10) = 7, put a bar over 9.
3. 8 – $\overline{9}$ = 8 – 10 = 8 + 0 (complementary of 10) = 8, put a bar over 8.
4. 9 – $\overline{8}$ = 9 – 9 = 0
5. 4 – 2 = 2

Example: Subtract 375428 from 88309

Solution:

$$\begin{array}{r} 3\ \ 7\ \ 5\ \ 4\ \ 2\ \ 8 \\ -\overline{0}\ \ \overline{8}\ \ 8\ \ 3\ \ \overline{0}\ \ 9 \\ \hline 2\ \ 8\ \ 7\ \ 1\ \ 1\ \ 9 \end{array}$$

Explanation:

1. $8 - 9 = 8 + 1$ (complementary of 9) = 9, place a bar over 0.
2. $2 - \overline{0} = 2 - 1 = 1$
3. $4 - 3 = 1$
4. $5 - 8 = 5 + 2$ (complementary of 8) = 7, place a bar over 8.
5. $7 - \overline{8} = 7 - 9 = 7 + 1$ (complementary of 9) = 8, place a bar over 0.
6. $3 - \overline{0} = 3 - 1 = 2$

Example: Subtract 3746735 – 2837546

Solution:

$$\begin{array}{r} 3\ 7\ 4\ 6\ 7\ 3\ 5 \\ -\overline{2}\ 8\ \overline{3}\ 7\ \overline{5}\ \overline{4}\ 6 \\ \hline 9\ 0\ 9\ 1\ 8\ 9 \end{array}$$

Explanation:

1. 5 – 6 is not possible with (+) sign, so take the complementary of 6 which is 4 and add it to 5. So $5 - 6 = 5 + 4 = 9$
2. Apply Ekadhikena Purvena sutras and place a bar over 4 (preceding 6) making it 1 richer than itself.

   i.e. $\overline{4} = 4 + 1 = 5$
3. Again $3 - \overline{4} = 3 - 5$ is not possible with positive sign, so we take the complementary of 5 and add it with 3. Hence $3 - \overline{4} = 3 + 5 = 8$. Now place a bar over 5 (digit preceding 4)
4. $7 - \overline{5} = 7 - 6 = 1$
5. 6 – 7 is again not possible with (+) sign, so write the complementary of 7 and add it with 6 i.e. $6 - 7 = 6 + 3 = 9$ and a bar is placed over the preceding digit making it 1 richer than itself.
6. $4 - \overline{3} = 4 - 4 = 0$
7. $7 - 8 = 7 + 2$ (complementary value of 8) = 9. Place a bar over 2.
8. Lastly $3 - \overline{2} = 3 - 3 = 0$

Hence, the whole process of subtraction gets completed.

Example: Subtract 48769248 from 19358769

Solution:

$$\begin{array}{r} 4\;8\;7\;6\;9\;2\;4\;8 \\ -1\;9\;3\;5\;8\;7\;6\;9 \\ \hline \end{array}$$

The entire process of subtraction has been illustrated here step by step with detailed explanation so that even a lay man can understand the whole process easily.

- 8 – 9 = 8 + 1 (complement of 9) = 9, place a bar over 6.
- 4 – $\overline{6}$ = 4 – 7 = 4 + 3 (complement of 7) = 7, place a bar over 7.
- 2 – $\overline{7}$ = 2 – 8 = 2 + 2 (complement of 8) = 4, place a bar over 8.
- 9 – $\overline{8}$ = 9 – 9 = 0
- 6 – 5 = 1
- 7 – 3 = 4
- 8 – 9 = 8 + 1 (complement of 9) = 9, place a bar over 1.
- 4 – $\overline{1}$ = 4 – 2 = 2

Since there are no more digits left in the minuend or subtrahend hence subtraction process gets completed. The whole process can be done in one line.

$$\begin{array}{r} 4\;8\;7\;6\;9\;2\;4\;8 \\ -\overline{1}\;9\;3\;5\;\overline{8}\;\overline{7}\;\overline{6}\;9 \\ \hline 2\;9\;4\;1\;0\;4\;7\;9 \end{array}$$

Ekanyuena Purvena and Complementary Method (एकन्यून पूर्वेण व परममित्र विधि):

The meaning of both the terms has been discussed in the beginning. In order to understand how this method works, let us take a few examples and see the *modus operandi*.

Example: Subtract 28 from 42

Solution:

$$\begin{array}{r} 4\;2 \\ -2\;8 \\ \hline 1\;4 \end{array}$$

Explanation:

1. 2 – 8 = – 6 i.e. it is 4 short of 10(10 – 4 = 6 by complementary method). Write 4 in the remainder column below 2 – 8.
2. Since there is a negative sign at unit place, so you have to reduce the preceding minuend (4) by 1 (Apply Ekanyuena Purvena). So 3 – 2 = 1 will be written below 4 – 2 in the remainder column.
3. Hence 42 – 28 = 14 is the correct answer.

Take another example for better understanding.

Example: Subtract 854 – 278

Solution:

```
  8 5 4
 -2 7 8
 ------
  5 7 6
```

Explanation:

1. 4 – 8 = – 4, i.e. it is 6 short of 10(10 – 6 = 4). Write 6 in the remainder column below 4 – 8
2. Since there is a negative (–) sign at unit place so you have to reduce the preceding minuend 5 by 1 (Use Ekanyuena Purvena). So the next minuend is 4 and the subtrahend is 7.
3. 4 – 7 = – 3. It is again 7 short of 10, so write 7 below the second column. The remainder again being negative will affect the value of minuend 8 by reducing it to 1. Thus the last minuend is 7.
4. 7 – 2 = 5

Example: Subtract 5782 from 2997

Solution:

```
  5 7 8 2
 -2 9 9 7
 --------
```

A detailed step-by-step method is given here for you.

1. 2 – 7 = - 5, so write 5 (complementary of 5) in the remainder column.
2. Since the unit difference is negative, the preceding minuend needs to be reduced by 1. Hence in place of 8 – 9 in the second column,

you will take 7 – 9 into consideration. Here, Ekanyuena Purvena Sutra has been used.

3. 7 – 9 = – 2, so write 8(complementary of –2) should be written in the remainder column.
4. Again, the minuend in the third column will be further reduced by 1. So 7 – 9 of third column wil! be treated as 6 – 9 in actual operation.
1. 6 – 9 = – 3, hence 7 (complementary of – 3) will be written in the remainder column.
1. The minuend of the fourth column, by the Ekanyuena Sutra, gets reduced by 1 and hence 4 – 2 = 2 is written in the remainder column.

The whole step can be summarised in one step as–

$$\begin{array}{r} 5\ 7\ 8\ 2 \\ -2\ 9\ 9\ 7 \\ \hline 2\ 7\ 8\ 5 \end{array}$$

In order to have a better understanding of this method let us take a few more examples and treat them differently this time. I have named this different method as **Digit Separator Method**. This method simply works on the very principle of Ekanyu ena Purvena but the presentation seems to be something different.

Digit Separator Method

The *modus operandi* of this method is explained in the example given below.

Example: Subtract 46 from 29

Solution:

- Draw as many vertical digit separator lines as the number of digits are.

1st	2nd
4	6
–2	9

- Whenever the minuend at the top is less than the subtrahend at the bottom, put (– 1) in the remainder column of each digit separator.

	1st	2nd	
	4	6	———— Minuend
–	2	9	————Subtrahend
		(–1)	

- Here 6 < 9 in the 2nd column, hence (–1) encircled is placed in the respective remainder column.
- Subtract algebraically, putting the sign of larger number in the remainder column.

1st	2nd	
4	6	———— Minuend
–2	9	————Subtrahend
2	(–1) –3	

- Write the complement of negative remainder. The complement of 3 is 7.

1st	2nd	
4	6	———— Minuend
–2	9	————Subtrahend
2	(–1) 7	

- Now subtract as the arrow moves from left to right.

2 (–1) 7 = 17

Example: Subtract 356 from 198

Solution:

- Draw as many vertical digit separator lines

1st	2nd	3rd
3	5	6
–1	9	8

- Whenever the minuend at the top is less than the subtrahend at the bottom, put (– 1) in the remainder column of each digit separator.

1st	2nd	3rd	
3	5	6	Minuend
–1	9	8	Subtrahend
			Remainder
	(–1)	(–1)	

Here 6 < 8 and 5 < 9, so the (–1) encircled is placed in 2nd and 3rd column.

- Subtract algebraically, putting the sign of larger number in the remainder column.

1st	2nd	3rd	
3	5	6	Minuend
–1	9	8	Subtrahend
2	–4	–2	Remainder
	(–1)	(–1)	

- Write the complement of negative remainder in the 2nd and 3rd column. Here the complement of –2 and –4 is 8 and 6.

1st	2nd	3rd	
3	5	6	Minuend
–1	9	8	Subtrahend
2	6	8	Remainder
	(–1)	(–1)	

- Now subtract from left to right as the arrow moves.

2 (–1) 6 (–1) 8 = 158

Example: Subtract 2894 from 4547.

Solution:

- Draw as many vertical digit separator lines as the number of digits are.

1st	2nd	3rd	4th
4	5	4	7
2	8	9	4

- Whenever the minuend at the top is less than the subtrahend at the bottom, put (–1) in the remainder column of each digit separator.

1st	2nd	3rd	4th	
4	5	4	7	———— Minuend
2	8	9	4	———— Subtrahend
				———— Remainder
	(–1)	(–1)		

Here, 4 < 9 and 5 < 8 in 3rd and 2nd column which means that minuend < subtrahend. Hence –1 encircled is placed in the respective remainder column.

- Now subtract as you do in algebra. i.e. put the sign of larger number in the remainder column e.g. 4 – 7 = – 3 . Since 7 is larger than 4 and 7 has a negative sign so we subtract 4 from 7 and put negative sign with the result.

1st	2nd	3rd	4th
4	5	4	7
2	8	9	4
2	–3	–5	3
	(–1)	(–1)	

- Write the complement of negative remainder written in the 2nd and 3rd column. Here the complement of –3 and –5 are respectively 7 and 5.

1st	2nd	3rd	4th
4	5	4	7
2	8	9	4
2	7	5	3
	(–1)	(–1)	

- Now subtract from right to left as the direction of arrow moves.

2 (–1) 7 (–1) 5 3 -----Remainder Column

= **1653.**

Hence

4547 – 2894 = 1653

Example: Subtract 375988 from 823479.

Solution:

- Draw digit separator vertical line.

1st	2nd	3rd	4th	5th	6th	
8	2	3	4	7	9	Minuend
–3	7	5	9	8	8	Subtrahend

- Subtract algebraically and write the remainder in the remainder column.

1st	2nd	3rd	4th	5th	6th	
8	2	3	4	7	9	Minuend
–3	7	5	9	8	8	Subtrahend
5	–5	–2	–5	–1	1	Remainder

- In each of the remainder column with negative sign, write –1 encircled. Here the 2nd, 3rd, 4th and 5th column has negative remainder.

1st	2nd	3rd	4th	5th	6th	
8	2	3	4	7	9	Minuend
–3	7	5	9	8	8	Subtrahend
5	–5	–2	–5	–1	1	Remainder
(–1)	(–1)	(–1)	(–1)			

- Write the complement of each digit with negative sign. Here the complement of –5, –2, –5, and –1 in the 2nd, 3rd, 4th and 5th column are respectively 5, 8, 5 and 9.

1st	2nd	3rd	4th	5th	6th	
8	2	3	4	7	9	Minuend
3	7	5	9	8	8	Subtrahend
5	5	8	5	9	1	Remainder
(–1)	(–1)	(–1)	(–1)			

- Subtract from left to right in the direction of arrows to get the required result. For the convenience of readers, the remainder column is written here with arrow sign.

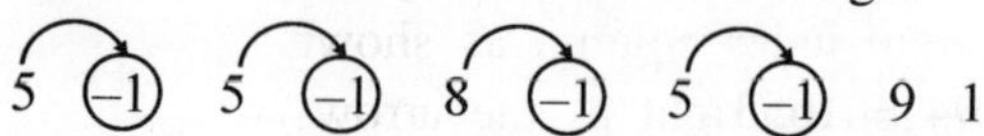

Hence

8	2	3	4	7	9
–3	7	5	9	8	8
4	4	7	4	9	1

I do hope that the readers have enjoyed all the Vedic methods for subtraction very much. Let us finish this chapter on Vedic subtraction with one more example on digit separator subtraction method.

Example: Subtract 57625 – 29848

Solution:

- Draw digit separator vertical lines and subtract algebraically, with the sign of larger digit in the remainder column.

1st	2nd	3rd	4th	5th	
5	7	6	2	5	Minuend
–2	9	8	4	8	Subtrahend
3	–2	–2	–2	–3	Remainder

- In column 2nd, 3rd, 4th, and 5th with negative remainder place – 1 encircled.

1st	2nd	3rd	4th	5th	
5	7	6	2	5	Minuend
–2	9	8	4	8	Subtrahend
3	–2	–2	–2	–3	Remainder
	(–1)	(–1)	(–1)	(–1)	

- Write the complement of each of the negative number in 2nd , 3rd, 4th, and 5th column. The complement of –2, –2, –2 and –3 are respectively 8, 8, 8, and 7.

1st	2nd	3rd	4th	5th	
5	7	6	2	5	Minuend
–2	9	8	4	8	Subtrahend
3	8	8	8	7	Remainder
	(–1)	(–1)	(–1)	(–1)	

- Arrange the remainder column as shown below and subtract as directed from left to right as the arrows move.

$$3 \; \textcircled{-1} \; 8 \; \textcircled{-1} \; 8 \; \textcircled{-1} \; 8 \; \textcircled{-1} \; 7$$

Hence the final result is **27777.**

Exercise for Practice

Exercise 2.1

Subtract the following by Ekadhikena Purvena and Complementary Method.

1. $\begin{array}{r} 2569347 \\ -1598635 \\ \hline \end{array}$

2. $\begin{array}{r} 987265452 \\ -579676797 \\ \hline \end{array}$

3. $\begin{array}{r} 1879568964763 25 \\ -14621098731567 2 \\ \hline \end{array}$

4. $\begin{array}{r} 3257987 \\ -1597680 \\ \hline \end{array}$

5. $\begin{array}{r} 4879 \\ -2548 \\ \hline \end{array}$

6. $\begin{array}{r} 67989 \\ -57897 \\ \hline \end{array}$

7. $\begin{array}{r} 389 \\ -193 \\ \hline \end{array}$

8. $\begin{array}{r} 567 \\ -279 \\ \hline \end{array}$

Exercise 2.2

Subtract the following by Ekanyuena Purvena and Complementary Method.

1. $\begin{array}{r} 54830289 \\ -36542031 \\ \hline \end{array}$

2. $\begin{array}{r} 64879264 \\ -28940367 \\ \hline \end{array}$

3. 5 9 8 0 2 3
−4 8 9 7 9 0

4. 5 8 3 5 7 9 7 6 8 9
−3 5 4 7 9 0 2 4 7 6

5. 6 5 8 9 9 7
−4 6 7 9 2 5

6. 8 9 7 6 5 4 3
−2 4 9 8 6 7 2

7. 8 7 3 2
−5 4 8 9

8. 3 6 7 9 6
− 9 3 6 7

Exercise 2.3

Subtract the following with Digit Separator Method.

1. 6 6 6
−4 8 2

2. 6 9 8 0
−5 7 9 8

3. 9 8 4 7 9 8 5 4 6
−6 8 9 7 9 7 9 7 9

4. 8 7 0 5 6 9 7 6 9
−8 2 4 8 9 3 9 8 7

5. 4 7 9 8 6 7
−3 6 4 7 8 0

6. 7 5 6 2 7 9
−2 9 8 7 0 5

7. 9 2 6 7 9 9 8 2 5
−5 4 9 8 9 9 9 9 9

8. 4 6 8 7 9 3
−3 5 9 7 0 2

❑❑❑

3

MULTIPLICATION

Introduction

Multiplication is one of the most commonly used mathematical operations. The basic fact of multiplication is to add. In the conventional method, we have a unique pattern of multiplication being followed for centuries.

```
      4 2 3 5 ——→ Multiplicand
×         2 5 ——→ Multiplier
  -----------
    2 1 1 7 5
  + 8 4 7 0 ×
  -----------
  1 0 5 8 7 5     ——→ Product
```

This old practice of multiplying numbers does not allow any experimentation, whereas there are various methods of multiplication in Vedic Mathematics. This process of current multiplication which is taught everywhere in the school has many disadvantages. The biggest being the cramming up of the multiplication tables up to 20. The best advantage of Vedic Multiplication technique over the present-day method is that here one has the benefit of applying different sutras according to the category it falls under. On the one hand, there is Urdhva-Tiryak sutra which is applicable to all types of multiplication; on the other hand there are certain specific rules applicable for certain other cases.

Vedic Sutras for Multiplication

There are several Vedic sutras for multiplication. Some sutras are specific and have limited applications, whereas some others are applicable to all types of multiplication. It is now for the student to think of and to weigh all the possible alternatives available and make up his mind for the simplest method he chooses and then to apply the same to get the answer. Some of the Vedic Sutras discussed in the present chapter are as follows:

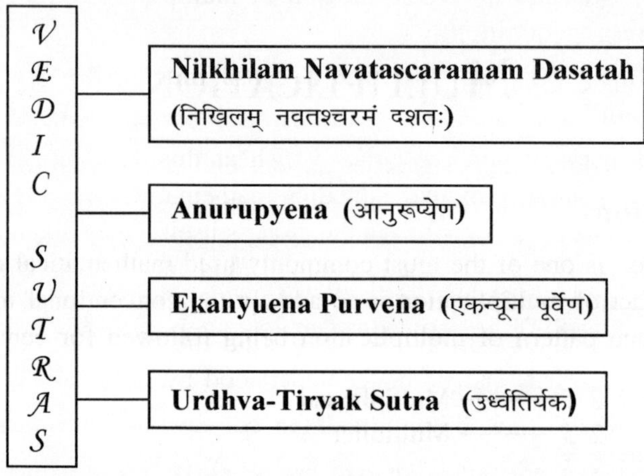

Meaning of Vedic Sutra

1. *Nikhilam Navatascaramam Dasatah* (निखिलम् नवतश्चरमं दशत:): It literally means **"All from nine and last from ten."**
2. *Anurupyena* (आनुरूप्येण): This Vedic sub-sutra simply means **"Proportionately"**. This sub-sutra is applicable when either the multiplicand or multipliers is sufficiently very far from the power of 10.
3. *Ekanyuena Purvena* (एकन्यून पूर्वेण): It literally means one less than the previous. This sutra has the limited application. It is used for multiplication wherein the multiplier digits consist entirely of nines. It comes up under three types.
   a. When the multiplicand and multipliers each consists of 2 digits.

b. When the multipliers have more digits than the multipliers.

c. When the multiplicand have more digits than the multipliers.

4. ***Urdhva-Tiryagbhyam*** (उर्ध्वतिर्यभ्याम्): It is a general formula applicable to all cases of multiplication. It is a process of vertical and cross-wise multiplication. This method has been further simplified and dealt with Dot and Cross method in this book.

Before I discuss the Vedic method of multiplication, let me put you a very interesting question.

Question: Why do you put a cross mark after each step of multiplication?

I think most of you are amazed to hear this question. But those of you who have grown up by understanding mathematics minutely can answer this question. The basic logic lies in the process itself. Once the multiplication of unit digit is over, we put a cross mark and move to find the product of ten's digit. Once the multiplication of ten's digit is over, we put a cross mark below the ten's digit and move to find the product of next digit at hundredth place. This can be better understood by the following example:

```
      3 4 5
    X 2 5 6
  ---------
    2 0 7 0
  1 7 2 5 X
+ 6 9 0 X X
  ---------
  8 8 3 2 0
```

Understand the basic structure of 256. In place value system 256 can be explicitly written as

$$256 = 2 \times 100 + 5 \times 10 + 6$$

Now look at the above multiplication and you will find that 5 of 256 is at the ten's place. Hence a cross is marked below the unit place and the final product of 345 × 5 begins from the ten's place. Similarly, 2 of 256 is at the hundred place and so is its product written from the hundred place and hence two cross marks one below the unit place and another below the tens place are made.

Vedic Multiplication

1. Nilkhilam Navatascaramam Dasatah Sutra (निखिलम् नवतश्चरमं दशत:)

First of all we shall discuss Nikhilam sutra for multiplication. The *modus operandi* of this sutra has been described here:

- Write the two numbers to be multiplied above and below at the right side of your copy.
- Take a base for your calculation. The base should be that power of 10, which is nearest to the number to be multiplied.
- Subtract each number from the base and write the remainder with respective plus (+) or minus (–) sign.
- The product will have two parts. Draw a vertical line for the demarcation.
- The left hand product is obtained by cross operation of two numbers written diagonally.
- The right hand product is obtained by multiplying the two digits written at the right hand portion.

Let us take a few examples to understand the step by step procedure to find the product of given number. We shall divide the whole procedure of multiplication in three cases.

CASE 1: When both the numbers to be multiplied are less than the base.

Example: Multiply 9 by 8.

Solution:

Step 1: $\begin{array}{r} 9 \\ \underline{\times\ 8} \end{array}$

Step 2: Both the numbers are nearer to the base 10, so Base = 10.

Subtract each number from the base 10 and write down the remainder with respective sign. The plus (+) sign is used to show that the number is greater than the base, whereas the minus (–) sign is used to show that the number is less than the base.

$$\begin{array}{r} 9 - 1 \\ \times\ 8 - 2 \\ \hline \end{array}$$

Step 3: Take any diagonal number and place the result to the left of the vertical line. The desired result after performing the operation is 7.

$$9 - 2 = 7$$

$$8 - 1 = 7$$

$$\begin{array}{r} 9 - 1 \\ \times\ 8 - 2 \\ \hline 7| \end{array}$$

Step 4: Multiply the two figures written at the extreme right. Put this number on the right side of the vertical line.

$(-1) \times (-2) = 2$

$$\begin{array}{r} 9 - 1 \\ \times\ 8 - 2 \\ \hline 7|\ 2 \end{array}$$

Hence 9 × 8 = 72 is the answer.

CASE 2: When both the numbers are above the base.

Example: Multiply 13 by 18.

Solution:

Step 1: Write the two numbers 13 and 18 above and below as shown here.

$$\begin{array}{r} 13 \\ \times\ 18 \\ \hline \end{array}$$

Step 2: Subtract the base from the number and write the difference against each number with respective sign.

$$13 - 10 = 3$$

$$18 - 10 = 8$$

Since both the numbers are greater than the working base 10 hence the plus (+) sign will be placed against the numbers.

$$\begin{array}{r} 13 + 3 \\ \times\ 18 + 8 \\ \hline \end{array}$$

Step 3: Take any diagonal and do the required operation. Write the result on the left side of the vertical line. Here 13 + 8 = 21 and 18 + 3 = 21.

$$\begin{array}{r|l} 13 + 8 & \\ \times\ 18 + 8 & \times \\ \hline 21 & \end{array}$$

Step 4: Multiply the right hand figure and write the product on the right side of the vertical line.

(+3) × (+ 8) = 24

$$\begin{array}{r} 13 + \ 3 \\ \times\ 18 + \ 8 \\ \hline 21 \mid 24 \end{array}$$

Now the question arises: **How to write the final answer?**

The answer is very simple. Since our base is 10, so the figure at the right side of the vertical line should be less than 10. In other words we can say that the right side of the vertical line should consist of only one digit. Let us conclude the answer by placing the surplus 2 of 24 to the left, as we do in simple multiplication.

$$\begin{array}{r} 13 + \ 3 \\ \times\ 18 + \ 8 \\ \hline 21 \mid 24 \\ + \end{array}$$

= 21 + 2 | 4

= 234

CASE 3: When one of the numbers is above the base and another is less than the base.

Example: Multiply 12 by 7.

Solution:

Step 1: Write the two numbers 12 and 7 above and below as shown here.

$$\begin{array}{r} 12 \\ \times\ 7 \\ \hline \end{array}$$

Step 2: Subtract the base from the number and write the difference against each number with respective sign.

12 – 10 = 2 and 7 – 10 = – 3

$$\begin{array}{r} 12 + 2 \\ \times\ 7 - 3 \\ \hline \end{array}$$

Step 3: Take any diagonal and do the required operation. Write the result on the left side of the vertical line.

12 – 3 = 9

Or 7 + 2 = 9

$$\begin{array}{r} 12 + 2 \\ \times\ 7 - 3 \\ \hline 9| \end{array}$$

Step 4: Write the product of extreme right figure on the other side of the vertical line.

(+2) × (– 3) = – 6

$$\begin{array}{r} 12 + 2 \\ \times\ 7 - 3 \\ \hline 9| - 6 \end{array}$$

When the right hand product is negative a Vinculum may be used i.e. $-6 = \overline{6}$.

Now the big question is **what next?**

When there is a minus (–) sign at the right hand product, use the Nikhilam formula which states "All from 9 and the last from 10."

Hence

$$\begin{array}{r|l} 12 & +\ 2 \\ 7 & -\ 3 \\ \hline = 9 & -\ 6 \\ = 8 & 10 - 6 \\ = 8 & \ 4 \end{array}$$

Therefore, $12 \times 7 = 84$.

Before taking a few more examples on all the three cases discussed so far, let us take some more examples which are nothing but the extension of the formula which says **"All from 9 and the last from 10."**

Example: Multiply 988 by 997

Solution:

Step 1: Here both the numbers are nearer to the working base 1000 so working base = 1000. Subtract both the numbers from the base

988 – 1000 = – 012 and 997 – 1000 = – 003

Step 2: In Vedic multiplication writing 12 and 3 instead of **012** or **003** will yield a wrong result.

$$\begin{array}{rr} 988 & -012 \\ \times 997 & -003 \\ \hline \end{array}$$

Step 3: Take any diagonal and perform the required operation. Write the result on the left side of the vertical line.

988 – 003 = 985

997 – 012 = 985

$$\begin{array}{rr} 988 & -012 \\ \times 977 & -003 \\ \hline 985| & \end{array}$$

Step 4: Multiply the right hand figure to obtain the product of the right side of the vertical line.

(– 012) × (– 003) = 036

$$\begin{array}{rr} 988 & -012 \\ \times 997 & -003 \\ \hline 985| & 036 \end{array}$$

A question may arise in your mind – **Why have I written 036 instead of 36?**

In order to understand it, let us take the following table into consideration.

Base	*Number of digits at the right side of vertical line*	
10	1	0
100	2	00
1000	3	000
10000	4	0000
100000	5	00000
1000000	6	000000
10000000	7	0000000

Table 1

On looking upon the above table, you can understand the basic fact that the figure at the right side of vertical line depends on the base you take. It is clear that for the purpose of convenience 12, 3 and 36 are written as 012, 003 and 036. The above table will help you to make the error minimum while multiplying for the figures to be placed at the right side of the vertical line.

Hence 988 × 997 = 985036

So far we have discussed all the three cases by taking one example of each type; now let us take a few more illustrations on each case to clear any more doubts.

ILLUSTRATION

Type 1

(Working Base = 10)

8 × 7 = ?	**9 × 1 = ?**	**6 × 8 = ?**
8 – 2	9 – 1	6 – 4
× 7 – 3	× 5 – 5	× 8 – 2
5 | 6	4 | 5	4 | 8
= 56	= 45	= 48

(Working Base = 100)

94 × 93 = ?
94 – 6
× 93 – 7
87 | 42
= 8742

98 × 65 = ?
98 – 2
× 65 – 35
63 | 70
= 6370

95 × 97 = ?
95 – 5
× 97 – 3
92 | 15
= 9215

88 × 99 = ?
88 – 12
× 99 – 1
87 | 12
= 8712

92 × 98 = ?
92 – 8
× 98 – 2
90 | 16
= 9016

89 × 96 = ?
89 – 11
× 96 – 4
85 | 44
= 8544

(Working Base = 1000)

888 × 997 = ?
888 – 112
× 997 – 3
885 | 336
= 885336

688 × 998 = ?
688 – 312
× 98 – 2
684 | 624
= 684624

698 × 997 = ?
698 – 302
× 997 – 3
695 | 906
= 695906

599 × 998 = ?
599 – 401
× 998 – 2
597 | 802
= 597802

993 × 975 = ?
993 – 7
× 975 – 25
968 | 175
= 968175

788 × 997 = ?
788 – 212
× 997 – 3
785 | 636
= 785636

You can enjoy the beauty of Vedic multiplication technique by taking a few more examples with base 10000, 100000… etc.

Type 2:

(Working Base = 100)

88 × 87 = ?
88 – 12
× 87 – 13
75 | 156
(+)
= 7656

82 × 92 = ?
82 – 18
× 92 – 8
74 | 144
(+)
= 7544

25 × 98 = ?
25 – 75
× 98 – 2
23 | 150
(+)
= 2450

105 × 192 = ?

$$\begin{array}{r|l} 105 + & 5 \\ \times 192 + & 92 \\ \hline 197 & 460 \end{array}$$

= 20160

108 × 124 = ?

$$\begin{array}{r|l} 108 + & 8 \\ \times 124 + & 24 \\ \hline 132 & 192 \end{array}$$

= 13392

115 × 107 = ?

$$\begin{array}{r|l} 115 + & 15 \\ \times 107 + & 7 \\ \hline 122 & 105 \end{array}$$

= 12305

(Working Base = 1000)

112 × 998 = ?

$$\begin{array}{r|l} 112 - & 888 \\ \times 998 - & 2 \\ \hline 110 & 1776 \end{array}$$

= 111776

884 × 991 = ?

$$\begin{array}{r|l} 884 - & 116 \\ \times 991 - & 9 \\ \hline 875 & 1044 \end{array}$$

= 876044

594 × 995 = ?

$$\begin{array}{r|l} 594 - & 406 \\ \times 995 - & 5 \\ \hline 589 & 2030 \end{array}$$

= 591030

When the product at the right hand side is more than the permissible digits, carry the surplus digit to the left as done in ordinary addition. For understanding the concept of permissible digits at the right side of the vertical line, the reader may see **Table 1**.

Type 3:

(Working Base = 100)

107 × 95 = ?

$$\begin{array}{r|l} 107 + & 7 \\ \times 95 - & 5 \\ \hline 102 & -35 \end{array}$$

= 102 | 100 – 35
= 10165

108 × 94 = ?

$$\begin{array}{r|l} 108 + & 8 \\ \times 94 - & 6 \\ \hline 102 & -48 \end{array}$$

= 101 | 100 – 48
= 10152

103 × 98 = ?

$$\begin{array}{r|l} 103 + & 3 \\ \times 98 - & 2 \\ \hline 101 & -2 \end{array}$$

= 100 | 100 – 6
= 10094

Whenever the right hand product is with the negative sign, subtract 1 from the left hand figure, and subtract the right hand figure from the base.

(Working Base = 1000)

1026 × 997 = ?

$$\begin{array}{r|l} 1026 + & 26 \\ \times 997 - & 3 \\ \hline 1023 & -078 \end{array}$$

= 1023 – 1 | 1000 – 078
= 1022922

1045 × 995 = ?

$$\begin{array}{r|l} 1045 + & 045 \\ \times 995 - & 005 \\ \hline 1040 & -225 \end{array}$$

= 1040 – 1 | 1000 – 225
= 1039775

1048 × 994 = ?

$$\begin{array}{r} 1048 + 048 \\ \times\ 994 - 006 \\ \hline 1042 \mid -288 \end{array}$$

= 1042 – 1 | 1000 – 288

= 1041712

1050 × 996 = ?

$$\begin{array}{r} 1050 + 050 \\ \times\ 996 - 004 \\ \hline 1046 \mid -200 \end{array}$$

= 1046 – 1 | 1000 – 200

= 1045800

The number of permissible digits at the right hand side should be decided on the basis of **Table 1.** If there is any deficiency in the right hand figure, zero should be placed at the left. Let us see this example to have a better picture of this rule.

Example: Multiply 9997 by 9998.

Solution:

$$\begin{array}{r} 9997 \\ \times 9998 \\ \hline \end{array}$$

Here the working base for both the numbers is 10000. According to the **Nikhilam** method described above, we have to subtract each number from the base and write the remainder with the plus (+) or minus (–) sign against the number.

9997 – 10000 = – 3

9998 – 10000 = – 2

$$\begin{array}{r} 9997 \quad -3 \\ \times 9998 \quad -2 \\ \hline 9995 \mid 0006 \end{array}$$

Since base = 10000, so the right hand side should have 4 digits at the maximum, hence 000 will be placed before 6 in order to obtain 4 permissible digits at the right.

Therefore,

$$\begin{array}{r} 9997 \quad -3 \\ \times 9998 \quad -2 \\ \hline 9995 \mid 0006 \end{array}$$

To avoid any such discrepancy, the readers are suggested to put the maximum permissible digits at the right side, in the very beginning. Let me take a few more examples to clarify my point.

(Working Base= 1000000)		(Working Base= 10000)	
999972	–000028	9992	– 0008
× 999998	–000002	× 9958	– 0042
999970	000056	9950	0336

In both the examples taken above, the number of permissible digits have been written at the very beginning to avoid any further confusion. In the first example the base is 1000000, so number of permissible digits at the right should have been 6 (see Table 1); hence 28 and 2 have been written as 000028 and 000002. On the other hand, the working base for the second example is taken as10000 and the difference from the base and the multiplying digits have been adjusted likewise as done in previous example.

The **Nikhilam** sutras discussed so far are applicable only when the multiplicand and multiplier are nearer to the working base. But the big question now arises – **What about the multiplication of two numbers, neither of which is near the convenient base?**

Let us understand this problem practically. Suppose we have to multiply 59 by 56. If we take 100 (a power of 10) as our base, then we shall get 41 and 44 as the deficiency from the base. Moreover, the consequent vertical multiplication of deficiency 41 and 44 would not serve the very purpose of Vedic multiplication.

Now the question is – **What next?**

The Anurupyena Sutra has the solution to such a problem.

2. Anurupyena Sutra (आनुरूप्येण)

The word Anurupyena simply means **"Proportionately."** This method is applicable only when the multiplicand and multipliers are very far from the theoretical base. In other words, when the multiplicand and multipliers are distant from the power of 10, we use this sutra.

Let us understand the *modus operandi* of Anurupyena sutra. We take two types of base: one is theoretical base and the other is working base. Theoretical base is taken in power of 10 and the working base is taken as the multiple of 10. Take this example to make is clear.

Sub-case 1: When the left hand figure is completely divisible.

Example:

$$\begin{array}{r} 48 \\ \times 42 \\ \hline \end{array}$$

Solution: Theoretical base = 100

Working base = 100 I 2 = 50

$$\begin{array}{r} 48-2 \\ \times 42-8 \\ \hline \end{array}$$

The working procedure is almost same except the fact that the left hand side figure will be divided by 2 as our working base is half of the theoretical base. As far as the right hand figure is concerned, it will remain unaffected.

$$\begin{array}{r|l} 48 \quad -2 & \\ 42 \quad -8 & \\ \hline 40 & 16 \end{array}$$

= ½ × 40 I 16

= 2016

Sub-case 2 : When the left hand figure on division gives a fractional quotient.

Example: 48 × 49 = ?

Solution: Theoretical base = 100 Working base = 50 = (100 ÷ 2)

Step 1: Writing the deficiency (–2) and (–1) against the number 48 and 49 taken from its working base.

48 – 50 = – 2 and 49 – 50 = –1

$$\begin{array}{r} 48-2 \\ \times 49-1 \\ \hline \end{array}$$

Step 2: Do the required operation by taking any diagonal and writing the result on the left side of the vertical line. Multiply the right hand figure vertically.

$$\begin{array}{r|l} 48 \quad -2 & \\ \times 49 \quad -1 & \\ \hline 47 & 02 \end{array}$$

Step 3: Divide the left hand figure by 2. Here 47 on dividing by 2 give us a fractional quotient i.e. 23½ .

Step 4: The fractional part ½ (i.e. ½ of the theoretical base 100 = 50) is taken over to the right hand side.

$$\begin{array}{rr} 48 & -2 \\ \times\ 49 & -1 \\ \hline 47 | & 02 \end{array}$$

$$= \frac{1}{2} \times 47 | 02 = 23\frac{1}{2} | 02$$

$$= 23 | 50 + 2 = 2352$$

Sub-case 3: Wheln the right hand vertical product is negative.

Example:

$$\begin{array}{r} 52 \\ \times 48 \\ \hline \end{array}$$

Solution: Theoretical base = 100

Working base = 50 = (100 ÷ 2)

As discussed earlier, the excess or the deficiency from the working base is written against the number.

52 = 50 + 2 and 48 = 50 – 2

$$\begin{array}{r} 52 + 2 \\ \times 48 - 2 \\ \hline \end{array}$$

Step 1: Perform the desired operation diagonally and vertically.

$$\begin{array}{rr} 52 & +2 \\ \times\ 48 & -2 \\ \hline 50 | & -4 \end{array}$$

Step 2: Divide the left hand figure by 2

$$\begin{array}{r} 52 + 2 \\ \times 48 - 2 \\ \hline 2)\overline{50 | -04} \\ = 25 | -04 \\ 24 | 96 \end{array}$$

When the right hand figure is negative, the "***Nikhilam Navatas Caramam Dasatah***" formula will be used. This simply directs us to subtract 1 from the left hand figure and to subtract the right hand figure from the theoretical base.

Instead of taking the power of 10 as the theoretical base and 50, 250, 500...as the working base, we may take only one base which falls nearer to that number, according to our convenience. The best part of this rule is that you do not need to divide the left hand figure and hence the chance of getting the fractional part is almost zero.

$$\begin{array}{r} 41 \\ \times 48 \\ \hline \end{array}$$ (Working base: 10 × 4 = 40)

$$\begin{array}{r} 71 \\ \times 87 \\ \hline \end{array}$$ (Working base: 10 × 8 = 80)

$$\begin{array}{r} 252 \\ \times 299 \\ \hline \end{array}$$ (Working base: 100 × 3 = 300)

$$\begin{array}{r} 687 \\ \times 695 \\ \hline \end{array}$$ (Working base: 100 × 7 = 700)

The *modus operandi* of such base is very simple. In order to get the final answer you have to multiply the left hand figure by the multiple of working base.

Example:

$$\begin{array}{rr} 41 & +1 \\ \times 48 & +8 \\ \hline 49 \mid 8 & \\ = 49 \mid 8 & \\ \times 4 & \\ \hline = 196 \mid 8 & \\ = 1968 & \end{array}$$

Example:

$$\begin{array}{rr} 71 & -9 \\ \times 87 & +7 \\ \hline 78 \mid -63 & \\ \times 8 & \\ \hline = 624 \mid -63 & \\ = 6240 & \\ -63 & \\ \hline = 6177 & \end{array}$$

Example:

$$\begin{array}{rr} 252 & -48 \\ \times 299 & -01 \\ \hline 251 \mid 48 & \\ \times 3 & \\ \hline = 753 \mid 48 & \\ = 75348 & \end{array}$$

Example:

$$\begin{array}{r|l} 687 & -013 \\ \times 695 & -005 \\ \hline 682 & 65 \\ \times 7 & \\ \hline = 4774 & 65 \\ = 477465 & \end{array}$$

I do hope you have understood the Vedic concept of multiplication. Though initially it takes some time to work fast with the help of Vedic Sutras, but once you are adapted to these, you may find it very useful.

Let us discuss one very interesting Vedic Sutra which will help you to multiply any number with repeatance of 9 (i.e. 9, 99, 999…) in one line. This one-line Vedic multiplication technique is like child's play.

Ekanyuena Purvena (एकन्यून पूर्वेण)

The term **Ekanyuena Purvena** (एकन्यून पूर्वेण) literally means one less than the previous. In order to focus its applicability in a wider context, I have divided it into three segments.

CASE 1: When the multiplicand and multipliers are of equal digits.

Example:

(*a*) $\begin{array}{r} 12 \\ \times 99 \\ \hline \end{array}$ (*b*) $\begin{array}{r} 123 \\ \times 999 \\ \hline \end{array}$ (*c*) $\begin{array}{r} 8765 \\ \times 9999 \\ \hline \end{array}$ (*d*) $\begin{array}{r} 896456 \\ \times 999999 \\ \hline \end{array}$

In all the above examples, the number of 9's in the multiplier is equal to that of the digits of multiplicand.

How does this method work?

It is a one-line multiplication technique where result is obtained mentally. The *modus operandi* is as follows:

- Subtract 1 from the multiplicand and put it on the left side of vertical line.
- The digit at the right hand side is obtained on subtracting the remainder at the left from the multipliers.

Example:

```
 4 3 6 5
×9 9 9 9
```

Solution:

- Subtract 1 from the multiplicand i.e. 4365 – 1 = 4364. Write the result obtained on the left side of vertical line.
- Subtract the result obtained above, from the multipliers, and write it on the right side of vertical line.

```
      4 3 6 5
     ×9 9 9 9
  ------------
   4365 | 9999
     –1 |–4364
  ------------
   4364   5635
```

Hence 4365 × 9999 = 43645635

Example:

```
 8 9 3 4 5 4
×9 9 9 9 9 9
```

Solution:

- Subtract 1 from the multiplicand i.e. 896454 – 1 = 896453
- Subtract 896453 from 999999 i.e. 999999 – 896453 = 103546

Hence

```
     8 9 6 4 5 4
    ×9 9 9 9 9 9
896453|03103541
  = 896453103546
```

Let us take some more examples.

```
  1 2 3               4 5 7 6 9
 ×9 9 9              ×9 9 9 9 9
1 2 3|9 9 9         4 5 7 6 9|  9 9 9 9 9
  –1|–122                  –1|– 4 5 7 6 8
 122| 877           4 5 7 6 9|  5 4 2 3 1
= 122877            = 4576854231
```

CASE 2: When the number of digits of multiplicand is less than that of multipliers.

Procedure: The same procedure will be adopted as done in **case 1**

Let us take a few examples:

```
  44          3548          13324
×999       ×999999       ×9999999
```

In all the above examples, the number of digits of multiplicand is less than that of the digits of multiplier.

Illustration:

Case 3: When the multipliers have more digits than that of multiplicand.

```
  44           3458             13324
 ×999        x 999999         ×99999999
44|999       3458|999999      13324|99999999
--1|-43        -1|  -3457        -1|   -13323
43|956       3457|996542      13323|99986676
=43956       =3457996542      =1332399986676
```

This case is a little bit different from the last two cases discussed so far under Ekanynena Purvena (एकन्यून पूर्वेण). In order to get the result you have to—

(a) Put as many zeros, as the numbers of 9's to the multiplicand.

(b) Subtract the original multiplicand from the figure obtained in step 1.

Let us discuss a few examples in detail.

Example: 1564
×99

Solution:

1564 has 4 digits and there are two 9's, so the very concept of case 3 will be taken into account.

```
 1564
×99
```

(1) Since there are two 9's, put two zeros at the end of 1564 making it 156400.

(2) Subtract 1564 (original multiplicand) from 156400

i.e.
$$\begin{array}{r} 156400 \\ -1564 \\ \hline 154836 \end{array}$$

Hence 1564 × 99 = 154836

Example:

$$\begin{array}{r} 783459 \\ \times 9999 \\ \hline \end{array}$$

Solution:

(1) Since there are four 9's, put four zeros at the end of 783459 making it 7834590000

(2) Subtract the original number 783459 from 7834590000.

$$\begin{array}{r} 7834590000 \\ -783459 \\ \hline 7833806541 \end{array}$$

Hence 783459 × 9999 = 7833806541

Urdhva Tiryagbhyam (उर्ध्वतिर्यभ्याम्)

So far we have discussed three Vedic Sutras, but all of them had limited applicability. We shall now proceed to deal with a general formula of multiplication which is applicable in all the cases. This sutra is widely known as **Urdhva Tiryak** Sutra which means **"Vertically and Crosswise".** The best feature of this method is:

(a) Once in practice, you can multiply any number of digits and obtain the result in one line.

(b) You are free to multiply from both the ends.

How does this method work?

The **Urdhva Tiryak** method of multiplication is based on the simple algebraic principle. A simple example will suffice to clarify the *modus operandi* thereof.

$$\begin{array}{l} cx + a \\ \underline{dx + b} \\ cdx^2 + dax \\ \qquad \underline{+ cbx + ab} \\ cdx^2 + (da + cb)x + ab \end{array}$$

The coefficient of $\mathbf{x^2}$ is obtained by vertical multiplication of **c** and **d**. The middle term i.e. the coefficient of **x** is obtained by the cross product of **d** and **a** and of **c** and **b** and the addition of the two products. The last term is obtained by the vertical multiplication of **a** and **b.**

Take an example to understand the fact clearly.

$14 \times 12 = 1 \times 1 \mid 1 \times 2 + 1 \times 4 \mid 2 \times 4$

$= 1 \mid 6 \mid 8 = 168.$

A further simplification can be understood by **Dot and Stick Method**.

Dot and Stick Method

Multiplication of 2-digit numbers

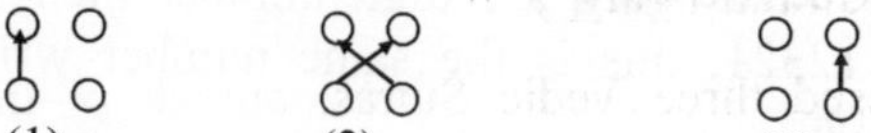

(1) (2) (1)

Multiplication of 3-digit number

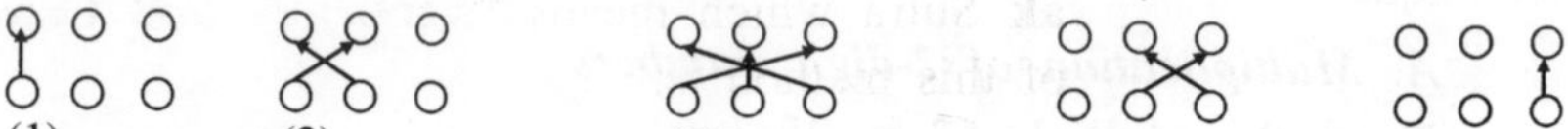

(1) (2) (3) (2) (1)

Multiplication of 4-digit number

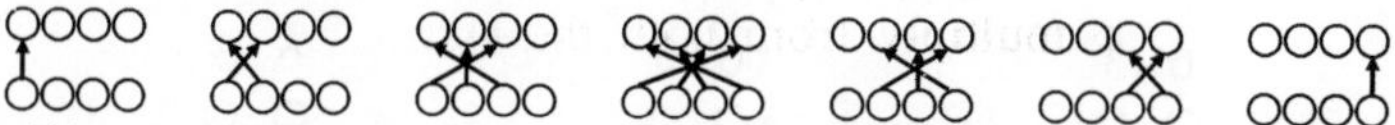

(1) (2) (3) (4) (3) (2) (1)

The dot and stick method may bring some amount of discomfort initially but the more you practice, the more comfortable you will feel. At the very beginning you might get puzzled over the multiplication by dot and cross as shown above pictorially, but here is an interesting technique to learn the dot and cross method. See this unique pattern of multiplication:

$11 \times 11 = 121$

$111 \times 111 = 12321$

$1111 \times 1111 = 1234321$

$11111 \times 11111 = 123454321$

Looking upon the multiplication of 11 × 11, 111 × 111 ...and the dot multiplication shown pictorially you may find a very special clue. Did you notice the clue?

If you are multiplying a two-digit number, see the product of 11 × 11 i.e. 121 and now notice the number written below the dots of 2-digit multiplication.

Did you find any similarity?

Yes, the same 121 is written there. This process can be summed up in three points :

- First multiply vertically the number placed at unit digits.
- Do the sum of cross product of two figures.
- Finally, multiply vertically the remaining digits.

Similarly, if you are multiplying a 3-digit number then look at the product of 111 × 111 = 12321, this is the same number written at the bottom of dot multiplication.

Let us take a few examples to clarify the dot and cross product technique.

A. Multiplication of 2-digit numbers

Example: Multiply 63 by 48

Solution:

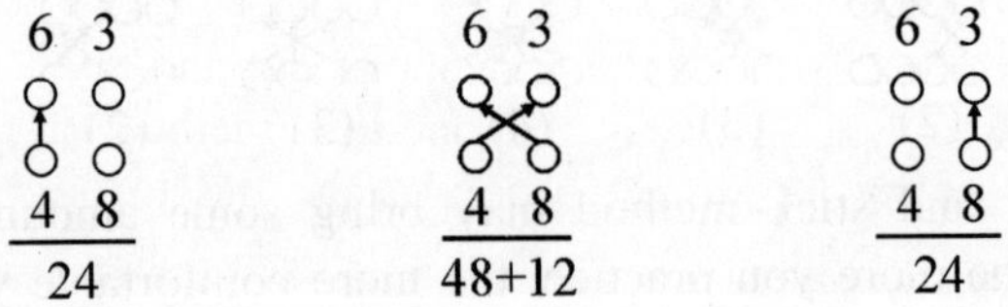

6 3	6 3	6 3
4 8	4 8	4 8
24	48+12	24

Arranging the number and adding them from right to left, taking only one digit at a time, we get the final result. -

= 2 4 | 6 0 | 2 4

= 3024

Example: Multiply 92 by 18 =?

Solution: Arranging the number on the dots.

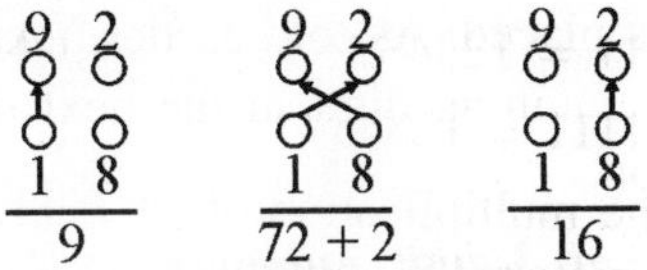

$$\begin{array}{ccc} 9\ 2 & 9\ 2 & 9\ 2 \\ 1\ 8 & 1\ 8 & 1\ 8 \\ \hline 9 & 72+2 & 16 \end{array}$$

Arranging the number and adding them from right to left, taking only one digit at a time, we get the final result.

$$= 9 \mid 7 4 \mid 1 6$$

$$= 1656$$

Example: Multiply 47 by 86

Solution: Arranging the number on the dots.

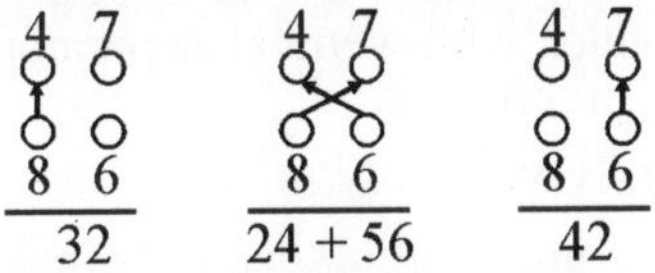

$$\begin{array}{ccc} 4\ 7 & 4\ 7 & 4\ 7 \\ 8\ 6 & 8\ 6 & 8\ 6 \\ \hline 32 & 24+56 & 42 \end{array}$$

Arranging the number and adding them from right to left, taking only one digit at a time, we get the final result.

$$= 3 2 \mid 8 0 \mid 4 2$$

$$= 4042$$

Once the concept is clear, the whole process can be done mentally in one line.

Example: Multiply 77 by 39.

Solution: I do hope the concept of dot and cross technique is clearly understood by you. Here is the one-line method. The sum of the cross multiplication of dots in the second stage has to be done mentally.

$$\begin{array}{l} \ \ 7 7 \\ \times 3 9 \\ \hline 2 1 \mid 8 4 \mid 6 3 \end{array}$$

$$= 3 0 0 3$$

In the 1st vertical separator 7 × 9 = 63 is written, in the second vertical separator the sum of the cross product of 7 × 9 and 3 × 7 is written directly i.e. 7 × 9 + 3 × 7 = 6 3 + 2 1= 8 4. In the third vertical separator 7 × 3 = 2 1 is placed. As told earlier, take only one digit in each separator and add the remaining digit to the next digit separator as shown with the arrow direction.

B. Multiplication of 3-digit numbers

Example: Multiply 548 by 159

Solution:

Arrange the number on the dots as shown below.

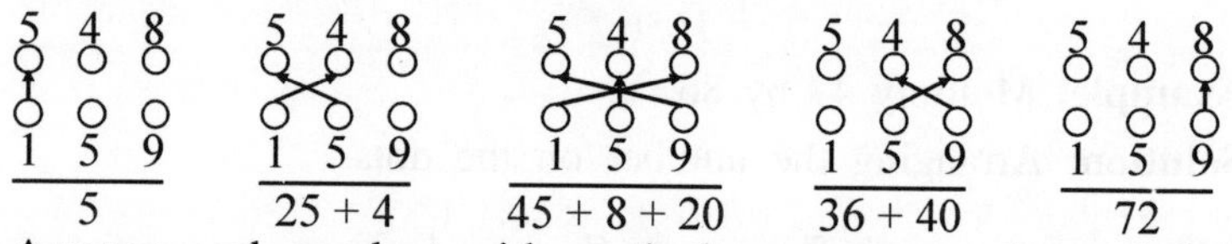

Arrange each product with vertical separator as shown below.

5 | 2 9 | 7 3 | 7 6 | 7 2

= 87132

Example: Multiply 659 by 898.

Solution: Arranging the numbers on the dots.

659 × 898	659 × 898	659 × 898	659 × 898	659 × 898
48	54 + 40	48 + 72 + 45	40 + 81	72

Arranging the product with vertical separator as shown below:

$$=48\,|\,94\,|\,165\,|\,121\,|\,72$$

$$=48\,|\,94\,|\,165\,|\,121+7\,|\,2$$

$$=48\,|\,94\,|\,165\,|\,\underline{12}\,8\,|\,2$$

$$=48\,|\,94\,|\,165+12\,|\,8\,|\,2$$

$$=48\,|\,94\,|\,\underline{17}\,7\,|\,8\,|\,2$$

$$=48\,|\,94+17\,|\,7\,|\,8\,|\,2$$

$$=48\,|\,\underline{11}\,1\,|\,7\,|\,8\,|\,2$$

$$=48+11\,|\,1\,|\,7\,|\,8\,|\,2$$

$$=59\,|\,1\,|\,7\,|\,8\,|\,2$$

$$=591782$$

Hence $659\times898=591782$

In the above example, the clear-cut detail has been provided for the better understanding of the whole process. In the next example, the readers are expected to do these operations involving the addition of two-three numbers of their own, and enjoy doing the calculation mentally. Once in running practice, you will notice that you don't need to write the dot and cross arrangement so many times and are able to perform the calculation involved in mid-step mentally.

Example: Multiply 247 by 989.

Solution: The whole operation of dot and cross method is done here in one line.

```
   2 4 7
 × 9 8 9
 ₁8 | ₅2 | ₁₁3 | ₉2 | ₆3
          8  2  3  2  3
      + 1  5 11  9  6
 ------------------------
       2  4  4  2  8  3
```

In the very beginning, I had mentioned the very fact that once the multiplication of every digit is completed, you have to add the digits from right to left, taking only one digits in each separator. Here the extra digits, leaving the unit digit in each separator, have been written in sub-script so that further addition becomes easy.

Example: Multiply 467 by 598.

Solution:

$$
\begin{array}{r}
467 \\
\times\, 598 \\
\hline
{}_{2}0 \mid {}_{6}6 \mid {}_{12}1 \mid {}_{11}1 \mid {}_{5}6 \\
0\ 6\ 1\ 1\ 6 \\
+\ 2\ \ 6\ 12\ 11\ 5 \\
\hline
2\ 7\ 9\ 2\ 6\ 6
\end{array}
$$

C. Multiplication of 4-digit numbers

After a detailed discussion on the dot and cross multiplication technique (an extract of the Vedic Sutra **Urdhva-Tiryak**), it is now expected from the readers that they can find the product of any number with the dot and cross method. The first example here is done in detail with each operation illustrated clearly, but the next example onwards is done in a single line with the hope that readers are very much acquainted with the process of multiplication.

Example: Multiply 6471 by 6212.

Solution: Arranging the numbers on the dots.

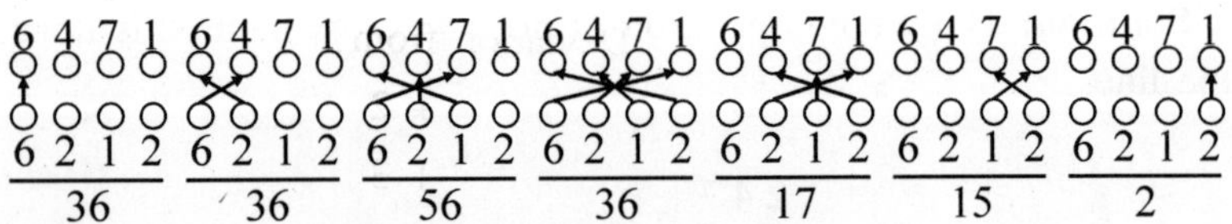

Arrange the numbers in vertical separators.

$$3\,6 \mid 3\,6 \mid 5\,6 \mid 3\,6 \mid 1\,7 \mid 1\,5 \mid 2$$

(+) (+) (+) (+) (+)

$$= 4\,0\,1\,9\,7\,8\,5\,2$$

Example: Multiply 8989 with 8892.

Solution:

```
 8989
×8892
64 | 136 | 208 | 233 | 162 | 97 | 18
= 6  4  6  8  3  2  7  8
   + 13 20 23 16  9  1
-------------------------------------
     7  9  9  3  0  1  8  8
```

Readers can extend the multiplication of 5, 6, or 7 digits by their own by dot and cross method. For multiplying two five-digit numbers, they can take help from the pattern of 11111 × 11111 = 123454321.

Multiplication of 3 numbers by Vedic method

Without wasting much time, the *modus operandi* of the whole process is illustrated with the help of some examples.

CASE 1: When the numbers to be multiplied are nearer to the power of 10.

Example: 12 × 13 × 15 = ?

Solution:

Step 1: Write the deviation of each number from its base. Place it against each number. For the above multiplication the working base = 10, and their deviations from the base are 2, 3 and 5 respectively.

Number	Deviation from the Base
12	+ 2
13	+ 3
15	+ 5

Step 2: Add all the deviations to the working base.

10 + 2 + 3 + 5 = 20

Step 3: Make all possible permutations of the deviation taking two at a time and add them. For example, the possible permutation of **abc** is **ab, bc** and **ac.**

2 × 3 + 2 × 5 + 3 × 5 = 31

Step 4: Multiply the deviations.

$2 \times 3 \times 5 = 30$

Step 5: Arrange the result obtained in above steps as shown here.

2 0 | 3 1 | 3 0

Adding the result from right to left in the direction of arrow, we can find the result.

20 | 31 | 30

= 2340

Example: $105 \times 104 \times 109 = ?$

Solution: The working base = 100

Step 1: Write the deviation of each number from its base. Place it against each number.

Number	Deviation
105	+ 5
104	+ 4
109	+ 9

Step 2: Find the sum of base and deviations.

Base + deviations = 100 + 5 + 4 + 9 = 118

Step 3: Multiply the deviations in pairs of two and sum up the result so obtained $5 \times 4 + 4 \times 9 + 5 \times 9 = 101$

Step 4: Multiply the deviations.

$5 \times 4 \times 9 = 180$

Step 5: Arrange the result of all the above steps in a vertical separator and add them up, from right to left as done previously. But never forget to keep two digits in each separator as constant because the base taken here is 100.

118 | 101 | 180

= 1190280

Example: $989 \times 995 \times 1012 = ?$

Solution: The working base for the above number = 1000

Step 1: Write the deviation of each number from its base. Place it against each number.

Number	Deviation
989	– 11
995	– 5
1012	+ 12

Step 2: Find the sum of base and deviations.

Base + deviations = 1000 – 11 – 5 + 12 = 996

Step 3: Multiply the deviations in pairs of two and sum up the result soobtained.

(–11) × (–5) + (–5 × 12 + (–11) × (+12)

= 5 5 – 6 0 – 1 3 2 = –137

Step 4: Multiply the deviations

–11 × – 5 × 12 = 660

Step 5: Arrange the result of all the above steps in a vertical separator.

= 9 9 6 | (– 1 3 7) | 6 6 0

= 9 9 5 | 1 0 0 0 – 1 3 7 | 6 6 0

= 9 9 5 | 8 6 3 | 6 6 0

= 9 9 5 8 6 3 6 6 0

CASE 2: When the working base is 50, 500, 5000...

The working rule of case 2 is the same as that of case 1 except the following:

- Write the last result in column.
- Keep the right hand figure of column 3 intact and divide the figure in the 2nd column by ½ and the figure in the 1st column by ¼.
- Add the digits of each column as shown in the example.

Example: 54 × 56 × 51 = ?

Solution:

Step 1: Write the deviation of each number from its base. Place it against each number.

Number	Deviation
54	+ 4
56	+ 6
51	+ 1

Step 2: Find the sum of base and deviations.

50 + 4 + 6 + 1 = 6 1

Step 3: Multiply the deviations in pairs of two and sum up the result so obtained.

4 × 6 + 4 × 1 + 6 × 1 = 34

Step 4: Multiply the deviations

4 × 6 × 1 = 24

Arrange the result obtained above in a vertical separator.

61 | 34 | 24

Make three columns and write the result accordingly as shown here.

Column 1	**Column 2**	**Column 3**
61	34	24
= 61 × ¼	34 × ½	24
= 15 + ¼	17	24
= 15	25 + 17	24
= 15 42 24		

Here the theoretical base = 100 and working base = 50 = 100 ÷ 2, so ½ = 50 and ¼ = 25 is taken.

Example: 54 × 48 × 61 = ?

Solution: The working base = 50 = 100 ÷ 2

Step 1: Write the deviation of each number from its base. Place it against each number.

Number	Deviation
54	+ 4
48	− 2
61	+ 11

Step 2: Find the sum of base and deviations.

Base + deviations = 50 + 4 − 2 + 11 = 63

Step 3: Multiply the deviations in pairs of two and sum up the result so obtained

$$4 \times (-2) + 4 \times 11 + 11 \times (-2) = -8 + 44 - 22 = 14$$

Step 4: Multiply the deviations.

$$4 \times -2 \times 11 = -88$$

The final result is 63 | 14 | – 88

Writing the result in columns

Column 1	Column 2	Column 3
$63 \times \frac{1}{4}$	$14 \times \frac{1}{2}$	– 88
$= 15 + \frac{3}{4}$	7	– 88
= 15	75 + 7	– 88
= 15	82	– 88
= 15	81	100 – 88
= 15	81	12

Hence $54 \times 48 \times 61 = 158112$

Multiplication using Napier's Rods

Multiplication can be done quite easily using Napier's rods. Before discussing how Napier's rods works, let us know about the life of Napier.

John Napier, a Scottish mathematician was born in 1550 in Edinburgh in Scotland. His father Archibald Napier was an important man of the 16th century. He is best known for his invention of logarithms. He also invented Napier's Rods.

Napier's rods or bones consist of ten oblong pieces of wood or other material with square ends. Each of the four faces of each rod contains multiples of one of the nine digits. The first rod containing the multiples of 0, 1, 9, 8, the second of 1, 2, 9, 7, the third of 0, 3, 6, 9, the fourth of 0, 4, 9, 5, the fifth of 1, 2 , 8, 7, the sixth of 1, 3, 8,6, the seventh of 1, 4, 8, 5, the eighth of 2, 3, 7, 6, the ninth of 2, 4, 7, 5 and the tenth of 3, 4, 6, 5. Each rod, therefore, contains on two of its faces multiples of digits that are complementary to those on the other two faces; and the multiples of a digit and its complement are reversed in position.

1	1	2	3	4	5	6	7	8	9
2	2	4	6	8	1/0	1/2	1/4	1/6	1/8
3	3	6	9	1/2	1/5	1/8	2/1	2/4	2/7
4	4	8	1/2	1/6	2/0	2/4	2/8	3/2	3/6
5	5	1/0	1/5	2/0	2/5	3/0	3/5	4/0	4/5
6	6	1/2	1/8	2/4	3/0	3/6	4/2	4/8	5/4
7	7	1/4	2/1	2/8	3/5	4/2	4/9	5/6	6/3
8	8	1/6	2/4	3/2	4/0	4/8	5/6	6/4	7/2
9	9	1/8	2/7	3/6	4/5	5/4	6/3	7/2	8/1

Napier's Rods are good for multiplying a long number by a single number but it can also be used to multiply a long number by another long number.

Example: Multiply 123456 by 7.

1	1	2	3	4	5	6
2	2	4	6	8	1/0	1/2
3	3	6	9	1/2	1/5	1/8
4	4	8	1/2	1/6	2/0	2/4
5	5	1/0	1/5	2/0	2/5	3/0
6	6	1/2	1/8	2/4	3/0	3/6
7	7	1/4	2/1	2/8	3/5	4/2
8	8	1/6	2/4	3/2	4/0	4/8
9	9	1/8	2/7	3/6	4/5	5/4

Arrange the rods as shown above and look at the numbers in the 7th row, starting from the right and write down the number obtained by adding the digits in the parallelogram as shown.

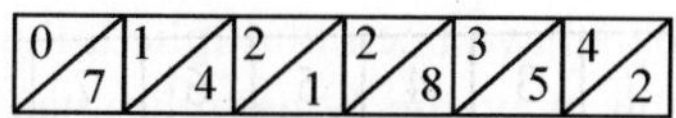

Now keep on adding the digits as discussed earlier in Vedic Multiplication.

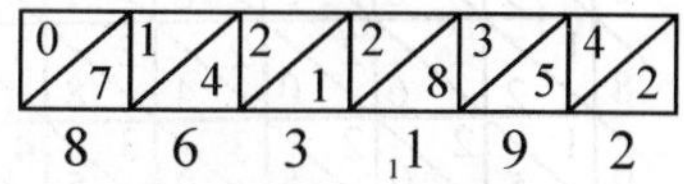

If the sum of digits in the respective column is more than 9, write the number exceeding in subscript as shown. Carry the subscript digit to add into the next column. Hence, the final sum will look like as shown here.

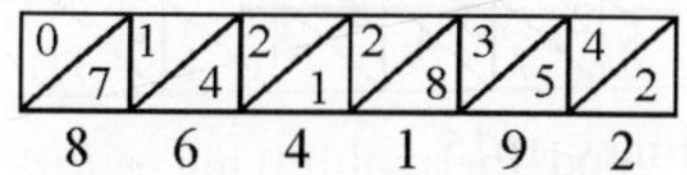

Hence the final result of multiplication is 864192.

Example: Multiply 46258 by 7.

Solution: The seventh square of rod 8 contains the number 56, so 56 is placed in the units position. The seventh square of rod 5 contains the number 35, so 35 is placed in the tens position. Continuing this process for the remaining rods, we get

```
4 × 7 = 2 8
6 × 7 =   4 2
2 × 7 =     1 4
5 × 7 =       3 5
8 × 7 =         5 6
---------------------
46258 × 7 = 3 2 3 8 0 6
```

Multiple-digit multipliers are processed the same way, digit by digit. These digits may be written down and shifted in a column just as in ordinary pencil and paper arithmetic.

Once all the numbers have been multiplied, the partial sums are added to find the result. In this way the bones eliminate the need to know or understand multiplication tables entirely, because multiplication has been replaced by a simple additive procedure.

Example: Multiply 1615 by 365.

Solution:

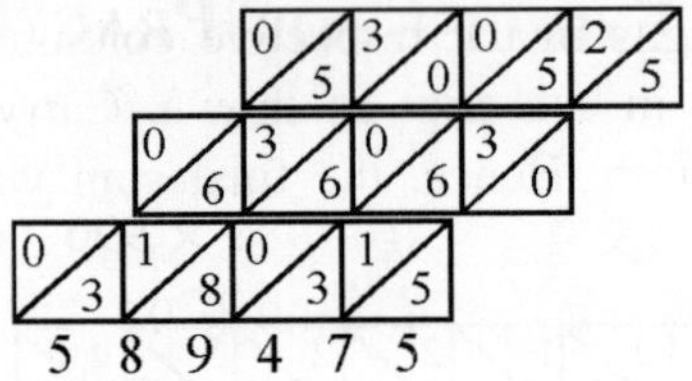

The result of 5 times 1615	=	8 0 7 5
The result of 6 times 1615 shifted one place left	=	9 6 9 0
The result of 3 times 1615 shifted two place left	=	4 8 4 5
		5 8 9 4 7 5

Hence $1615 \times 365 = 589475$

Example: Multiply 456 by 3579.

Solution:

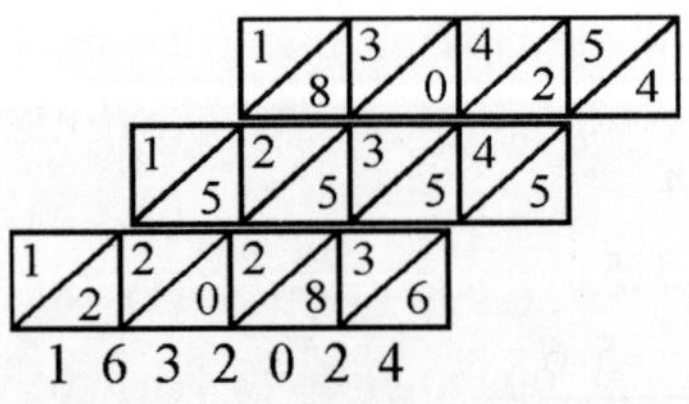

The result of 6 times 3579	=	2 1 4 7 4
The result of 5 times 3579 shifted one place left	=	1 7 8 9 5
The result of 4 times 3579 shifted two place left	=	1 4 3 1 6

Partial sum together = 2 1 4 7 4

$$\begin{array}{r} 21474 \\ 17895 \\ +\ 14316 \\ \hline 1632024 \end{array}$$

Hence 456 × 3579 = 1632024

EXERCISE FOR PRACTICE

Exercise 3.1

(a) 15 × 18 (b) 988 × 996 (c) 132 × 105

(d) 984 × 992 (e) 69 × 97 (f) 102 × 96

(g) 95 × 98 (h) 696 × 989 (i) 25 × 97

(j) 115 × 996 (k) 887 × 996 (l) 1024 × 996

(m) 1049 × 998 (n) 9996 × 9995 (o) 9999 × 9988

Exercise 3.2

Multiply the following using Anurupyena Sutra.

(a) 44 × 49 (b) 685 × 698 (c) 54 × 59

(d) 254 × 295 (e) 687 × 697 (f) 689 × 702

Exercise 3.3

Multiply the following using Ekanyuena Purvena Sutra.

(a) 854 × 999 (b) 652 × 999

(c) 3647 × 9999 (d) 548976 × 999999

(e) 65297458 × 99999999 (f) 459 × 99999

(g) 56489 × 999999999 (h) 65 × 999999

(i) 2546 × 999 (j) 564 × 99

(k) 45689 × 999 (l) 654 × 99999

Exercise 3.4

Multiply the following using Dot and Stick Method.

(a) 96 × 36	(b) 654 × 867	(c) 98 × 63
(d) 9658 × 3654	(e) 87325 × 56247	(f) 532 × 693
(g) 777 × 623	(h) 5697 × 6253	(i) 632 × 937
(j) 2574 × 9327	(k) 5632 × 3589	(l) 689 × 394
(m) 95672 × 35971	(n) 532 × 574	

Exercise 3.5

Multiply the following using Napier's Rods.

(a) 45632 × 8	(b) 543987 × 7	(c) 254 × 6
(d) 3256 × 65	(e) 5674 × 456	(f) 8732 × 653

❑❑❑

DIVISION

Introduction

Division is the reverse of multiplication just as subtraction is the reverse of addition. The method of division, taught in our school is somehow unacceptable to the kids. The method is itself so complicated that the student fears solving a problem of division. The main difficult part of division is that a student has to multiply long/big numbers by the trial digit of the quotient at every step and subtract that result from each dividend at each step. This is really a boring process and the main reason of unpopularity among the students.

The three Vedic sutras which we shall deal with in this chapter are really astonishingly lovely in the sense that the utmost multiplication one has to perform is of $9 \times 9 = 81$.

In Nikhilam method, there is no subtraction to be done at all. The second method Paravartya Yojyet is based on simple algebraic division and last method popularly known as Dhwajanka method of Division is based on long established Vedic process of mathematical calculations.

Different from the Nikhilam and Paravartya method, Dhwajanka method is capable of immediate application to all cases and it can be described as the "crowning gem of all' for the universality of its application.

There is a well-known and commonly used formula in division.

Dividend = Divisor × Quotient + Remainder

Let us understand the common division process with the following example.

```
               (DIVIDENT)
        123  )   4 8 5 2  (  39
  (DIVISOR)  )  -3 6 9    (  (QUOTIENT)
                 1 1 6 2
                -1 1 0 7
                     5 5 (REMAINDER)
```

Vedic Sutras for Division

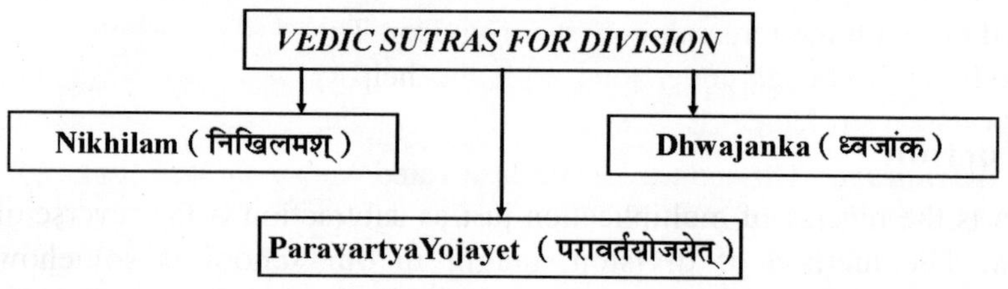

Meaning and Application of Vedic Sutra

Nikhilam (निखिलम्)**:** This method of division is based on the Nikhilam sutra of subtraction. This sutra is useful when every digit of divisor is greater than 5.

Advantage:

1. The best part of this Vedic sutra is that there is no subtraction to be done at all.
2. There is no problem of multiplying the divisor by trial digit, because the utmost multiplication one has to perform is of 9×9
3. The bigger the digits, the smaller will be its complement from the nearest base; and the multiplication task is lightened thereby.

Paravartya Yojyet (परावर्तयोजयेत्)**:** The basic meaning of this sutra is " Transpose and Apply." The well-known rule relating to transposition is the invariable change of sign with every change of side. This method like the Nikhilam method has limited application. In other words, when the digits in the divisor are small, the Nikhilam method is generally unsuitable, and the Paravartya one is always to be preferred.

Advantage:

1. The best advantage of this method is that you do not need to find the complement of divisor from the base.
2. The rest operation of division is done like a simple algebraic addition.

Dhwajanka (ध्वजांक)**:** The meaning is self-explanatory. The word Dhwaj means 'flag' and Anka means 'digit'. Hence, the literal meaning of this sutra is to place a digit or two of divisor at a top and division is then carried out with the remaining digits of divisor. The *modus operandi* of this method can be better understood with the help of some example, which will be taken up later.

Advantage: This is one of the best rated Vedic method for division, as it has universal applicability.

Nikhilam (निखिलम्) Method

The big question is - How to write the problem?

- Make three columns.
- The first column from the left is the divisor column.
- The second column from the left is the quotient column and
- The third column is the remainder column.

Working Rule:

1. Take a base (in the power of 10) nearest to the divisor and write its complement below the original divisor in the divisor column. Complement = Base – Divisor
2. The number of digits to be placed in the remainder column should be equal to the number of zeros in the base.
3. Carry down the first digit of dividend in the second column. This gives you the first digit of the quotient. Multiply the quotient digit by the complement, and place it in the dividend column; next to the first digit of dividend.

4. Write mechanically the sum of digit of second column to get the second digit of the quotient.
5. Repeat the process until you get a number in the remainder column, less than that of the original divisor.

Let us take some more examples to understand the *modus operandi* of this method.

Example: Divide 10025 by 88

Solution: Since every digit of divisor is more than 5, hence Nikhilam Sutra is applicable in this case.

Base = 100

Complement = 100 – 88 = 12

Column 1	Column 2 (Q)	Column 3 (R)
8 8	1 0 0	2 5
C = 1 2	1 2	- -
(Divisor)	1	2 -
	↓	3 6
	1 1 3	8 1

Hence,

Quotient = 113

Remainder = 81

Explanation:

1. Here Divisor = 88

   Nearest base in power of 10 = 100

   Complement = Base – Divisor

   = 100 – 88 = 12
2. Arrange digits in Column 2 and Column 3 accordingly. Since the base 100 has two zeros, so Column 3(Remainder column) should contain 2 digits. **(Rule 2)**
3. Carry 1 of column 2 (the first digit of dividend) down. This will give you the first digit of quotient.
4. Multiply the first digit of quotient with the complement and place it in the dividend column, next to the first digit of dividend.

   12 × 1 = 12 is placed below 0.

Column 1	Column 2 (Q)	Column 3 (R)
8 8	1 0 0	2 5
C = 1 2	1 2	- -
(Divisor)		
	1 1	

5. Carry down the sum of circled digits; this will give you the second digit of quotient. The second digit of quotient is 0 + 1 = 1
6. Multiply the second digit of quotient with the complement and place it in the dividend column, next to the second digit of dividend. 12 × 1 = 12 is placed below the second zero.

Column 1	Column 2 (Q)	Column 3 (R)
8 8	1 0 0	2 5
C = 1 2	1 2	- -
(Divisor)	1	2
	1 1 3	

7. Carry down the sum of second circled digits; this will give you the third digit of quotient. The third digit of quotient is 0 + 2 + 1 = 3
8. Multiply the complement 12 with the third quotient (3) and write it below the fourth digit of the remainder.
   12 × 3 = 36 is placed below 2 in the column 3.

Column 1	Column 2 (Q)	Column 3 (R)
8 8	1 0 0	2 5
C = 1 2	1 2	- -
(Divisor)	1	2 -
		3 6
	1 1 3	8 1

9. Sum up the digits of the Column 3 to get the Remainder. The above process repeats until the digits thus obtained in Remainder column is less than the original divisor.

   Here the sum of digits of Column 3 = 2 5 + 2 + 3 6 = 81

   And 81< 88 (Divisor), thus the process stops.

Example: Divide 11111111 by 99979

Solution: Base = 100000

Complement = 10 0 0 0 0 – 9 9 9 7 9 = 0 0 0 2 1

Since there are five zeros in the base, so the number of digits in the remainder column should be 5.

Arrange the dividend in Column 2 and Column 3 in such a way that the number of digits in Column 3 is 5. The first example has been explained in detail and hence the readers are expected to follow the same pattern in this example too.

Column 1	Column 2 (Q)	Column 3 (R)	
9 9 9 7 9	1 1 1	1 1 1 1 1	
C = 0 0 0 2 1	0 0	0 2 1 - -	**(00021 × 1)**
	0	0 0 2 1 -	**(00021 × 1)**
		0 0 0 2 1	**(00021 × 1)**
	1 1 1	1 3 4 4 2	

Hence, Quotient = 111

Remainder = 13442

Example: Divide 210012 by 8997

Solution: Base = 10000

Complement = 10000- 8997 = 1003

Number of digits to be placed in the remainder column = 4

	Column 1	Column 2 (Q)	Column 3 (R)
	8 9 9 7	2 1	0 0 1 2
C =	1 0 0 3	2	0 0 6 -
			3 0 0 9
		2 3	3 0 8 1

Quotient = 23

Remainder = 3081

Let us take another case when the remainder exceeds the divisor.

In case the remainder is greater than the divisor, divide the remainder by the original divisor, add the complement to the remainder.

Example: Divide 198 by 88

Solution: Base = 100

Complement = 100 – 88 = 12

Number of digits to be placed in remainder column = 2

Follow the same method of calculation until you get the **Remainder> Dividend**

	Column 1	Column 2 (Q)	Column 3 (R)	
	8 8	1	9 8	
C =	1 2		1 2	**(12 × 1 = 12)**
		1	**110**	

Since the remainder 110 > 88

Add the complement 12 in the remainder column

Column 1			Column 2 (Q)	Column 3 (R)		
	8	8	1	9	8	
C =	1	2		1	2	
			1	1	1	0
				+	1	2
			1	**1**	2	2

Q = 2 (1 + 1) **R = 2 2**

We add the newly obtained 1 to the previously obtained 1 and put down 2 as the quotient and 22 as remainder.

Example: Divide 1294567 by 89997

Solution: Base =100000

Complement = 100000 – 89997 = 10003

Number of digits to be placed in the remainder column = 5

Column 1	Column 2 (Q)		Column 3 (R)	
8 9 9 9 7	1	2	9 4 5 6 7	
1 0 0 0 3		1	0 0 0 3 –	**(10003 × 1)**
			3 0 0 0 9	**(10003 × 3)**
	1	3	1 2 4 6 0 6	
				(R = 124606 > 89997)
			+ 1 0 0 0 3	
	1	3	1 3 4 6 0 9	
	1	**4**	**3 4 6 0 9**	

Hence Quotient = 14 and Remainder = 34609

Example: Divide 11199171 by 99979

Solution: Base = 100000

Complement = 00021

Number of digits to be placed in Remainder column = 5

Column 1	Column 2 (Q)	Column 3 (R)
9 9 9 7 9	1 1 1	9 9 1 7 1
	0 0	0 2 1 - -
	0	0 0 2 1 –
		0 0 0 2 1
	1 1 1	1 0 1 5 0 2
		+ 0 0 0 2 1
	1 1 1	1 0 1 5 2 3
	1 1 2	0 1 5 2 3

Hence, Quotient = 112

Remainder = 01523

Paravartya Yojyet (परावर्तयोजयेत्)

This sutra is applicable only when the divisor is nearer to the base 10 or the power of 10. This sutra works effectively when the first digit of the divisor is 1.

Working Rule:

The working of this sutra is similar to the Nikhilam sutra except for the following:

1. In the divisor column, write the complement of divisor from the base and also the complement with change of sign. Thus + becomes – and conversely.
2. Complement = Divisor – Base
3. Here the multiplication of quotient digits is performed with the complement, having changed sign for each digit.

Example: Divide 1358 by 113

Solution: Base = 100

Complement = 113 – 100 = 13

Complement with changed sign for each digit = – 1 – 3

Column 1		Column 2 (Q)		Column 3 (R)	
	1 1 3	1	3	5	8
Complement =	1 3		–1	– 3	
Revised Complement =	–1–3			– 2	– 6
		1	**2**	**0**	**2**

Quotient = 12

Remainder = 02

Explanation:

1. Base = 100

   Complement = 113 – 100 = 13

   Writing each digit of complement with the changed sign = – 1 – 3. Since the number of zeros in the base is 2, so the number of digits in the Remainder column is 2.

2. Arrange the Dividend 1358 in Column 2 and Column 3 in such a way that the remainder contains a two – digit number.

Column 1		Column 2 (Q)		Column 3 (R)	
	1 1 3	1	3	5	8
Complement =	1 3				
Revised Complement =	–1 – 3				

3. Carry 1**(the fist digit of Column2)** down, this will give you the first digit of quotient.

Column 1		Column 2 (Q)		Column 3 (R)	
	1 1 3	1 ↓	3	5	8
Complement =	1 3				
Revised Complement =	–1–3				
		1			

4. Multiply the first digit of quotient with the **Revised complement** and write it next to the first digit of dividend.

(–1 – 3) × 1 = – 1 – 3

Now, add each digit of Column 2 and Column 3 to get the Quotient and Remainder.

Column 1	Column 2 (Q)		Column 3 (R)	
1 1 3	1	3	5	8
Complement = 1 3		–1	–3	
Revised Complement –1 –3	+		–2	–6
	1	**2**	**0**	**2**

Multiply the second digit of quotient with Revised Complement:

Example: Divide 395166 by 1321

Solution: Base = 1000

Complement = 1321 – 1000 = 321

Complement with changed sign for each digit = – 3 – 2 – 1

Column 1	Column 2 (Q)	Column 3 (R)	
1 3 2 1	3 9 5	1 6 6	
Complement 3 2 1	– 9 – 6	– 3 - -	(***)
Changed Complement			
–3–2 –1	0	0 0	(**)
		+3 +2+1	(*)
	3 0 -1	1 8 7	

*** The changed complement multiplied by 3(1st digit of the quotient)

** The changed complement multiplied by 0 (2nd digit of the quotient)

* The changed complement multiplied by (–1) (3rd digit of the quotient)

Quotient = 30 – 1 = 300 – 1 = 299

Remainder = 187

Dhwajanka (ध्वजांक) Method

The Dhwajanka method is quite helpful in dividing any number comfortably. The main divisor and the dhwajanka may be of as many digits as we want, depending upon our ability to perform the operation. We may take one digit or two digits as a dhwajanka.

Example: Divide 38982 by 73.

Solution:

The first step in Dhwajanka method is to make the column as shown below.

$$\underline{7^3|3898|2}$$

Step 1: Out of the divisor 73, we put down only the first digit i.e. 7 in the divisor column and put the other digit i.e. 3 on the top of the flag.

Here main divisor =7

Dhwajanka =3(one digit)

Hence, a line is drawn vertically from right leaving one digit at the end in the remainder column as shown above. Now the entire division is to be carried out by 7.

Step 2: As the first digit from the left of dividend (3) is less than 7, we take 38 as our first dividend. Divide 38 by 7

$\mathbf{Q_1=5}$ $\mathbf{R_1=3}$

$\mathbf{Q_1}$ = The first quotient

$\mathbf{R_1}$ = The first remainder.

Put the quotient below the horizontal line and prefix the remainder 3 below the digit 9.

$$\begin{array}{l} \underline{7^3|3\,8\,{}_39\,8|2} \\ \quad\quad\;\; 5 \end{array}$$

Step 3:

New gross dividend = 39

Subtract the product of dhwajanka 3 and the first quotient (5) i.e. 3 × 5 =15 from 39.

Net divided = 39 – 3 × 5 = 24

Divide 24 by 7

Q_2=3 **R_2=3**

Q_2 =The second quotient

R_2 = The second remainder

Prefix the remainder 3 below 8 above the horizontal line

$$\begin{array}{r|l|l} 7^3 & 3\,8\;{}_3 9\;{}_3 8 & 2 \\ \hline & \;\;\;53 & \end{array}$$

Step 4:

New gross dividend =38

Subtract the product of 2nd quotient and dhwajanka (3 × 3 = 9) from 38

i.e. 38 – 3 × 3 = 29

Net dividend = 29

Divide it by 7

Q_3 = 4 **R_3 = 1**

Q_3 = The third digit of quotient

R_3 = The third digit of remainder

Place 4 and 1 as discussed above.

$$\begin{array}{r|l|l} 7^3 & 3\,8\;{}_3 9\;{}_3 8 & {}_1 2 \\ \hline & \;\;\;534 & \end{array}$$

Step 5:

New gross dividend =12

$$\begin{array}{r|l|l} 7^3 & 3\,8\;{}_3 9\;{}_3 8 & {}_1 2 \\ \hline & \;\;\;534 & 0 \end{array}$$

Subtract the product of dhwajanka and third quotient (3 × 4 = 12) from 12

Remainder = 12 – 3 × 4 =0

Hence, **Quotient = 534** and **Remainder =0**

Example: Divide 98765 by 87

Solution: Writing the problem as stated above.

$$8^7 \,|\, 9\,8\,7\,6 \,|\, 5$$

Step 1: Main divisor = 8 Dhwajanka = 7

Step 2: Divide 9 by 8

$$\mathbf{Q_1=1 \;\; R_1=1}$$

Write the quotient below the horizontal line and prefix remainder below 8.

$$\frac{8^7 \,|\, 9\;{}_1 8\;7\;6 \,|\, 5}{1}$$

Step 3: Gross dividend =18

Subtract 7(Dhwajanka) × 1 (Q_1) = 7 from 18

Net dividend= 18 – 7 = 11

Divide 11 by 8

$$\mathbf{Q_2=1 \;\; R_2=3}$$

Write the quotient below the horizontal line and prefix remainder below 7

$$\frac{8^7 \,|\, 9\;{}_1 8\;{}_3 7\;6 \,|\, 5}{11}$$

Step 4. Gross dividend =37

Subtract product of Dhwajanka and Q_2 from 37 i.e. 37 – (7 × 1) =30

Divide 30 by 8

$$\mathbf{Q_3 = 3 \;\; R_3 = 6}$$

Place quotient and remainder in their respective place.

$$\frac{8^7 \,|\, 9\;{}_1 8\;{}_3 7\;{}_6 6 \,|\, {}_5 5}{113}$$

Step 5: Gross dividend = 66

Net dividend = 66 – 7 × 3 = 66 – 21 = 45

Divide 45 by 8

$$\mathbf{Q_4 = 5 \text{ and } R_4 = 5}$$

$$\begin{array}{l|l|l} 8^7 & 9\ {}_18\ {}_37\ {}_66 & {}_55 \\ \hline & 1135 & 20 \end{array}$$

Step 6: Gross dividend = 55

Remainder = 55 – 7 × 5 = 20

Hence, Quotient = 1135 and Remainder = 20

CASE 2: When Dhwajanka has more digits than the main divisor.

Example: Divide 72 38 761 by 524

Solution: If we take two digits as dhwajanka, then the remainder column will certainly have two digits. See the arrangement below.

$$\begin{array}{l|l|l} 5^{24} & 72\ 387 & 61 \\ \hline \end{array}$$

Step 1: Divide 7 by 5

Q_1 = 1 and R_1 = 2

$$\begin{array}{l|l|l} 5^{24} & 7\ 2\ 3\ 8\ 7 & 61 \\ \hline \end{array}$$

Step 2: Gross dividend =22

Net dividend= 22 -1 × 2 = 20

[The product of (Q_1 = 1) and first flag digits (2) is subtracted]

Step 3: Divide 20 by 5

Q_2 = 4 and R_2 = 0

Since R_2 = 0 so we have to take the quotient below 4 i.e. New quotient = 4 – 1 = 3. This is because we cannot consider the remainder zero in the middle of division. The same process will be applicable in case the net divisor in the middle is either **zero or negative**.

Hence, **Revised Quotient Q_2 = 3 and R_2 = 5**

$$\begin{array}{l|l|l} 5^{24} & 7\ {}_22\ {}_53\ 8\ 7 & 61 \\ \hline & 13 & \end{array}$$

Step 4: Gross dividend =53

Net dividend = 53 – (sum of cross product of two flag digits and two quotients)

= 53 – (2 × 3 + 1 × 4)

= 43

Divide 43 by 5

$Q_3 = 8$ and $R_3 = 3$

$$\begin{array}{r|l|l} 5^{24} & 7_2\ 2_5\ 3_3\ 8\ 7 & 61 \\ \hline & 138 & \end{array}$$

Step 5: Gross dividend =38

Net dividend = 38 – (sum of cross product of 38 and 24)

= 38 – (3 × 4 + 2 × 8)

= 38 – 28 = 10

Divide 10 by 5

$Q_4 = 2$ and $R_4 = 0$

Since $R_4 = 0$ so we have to take $Q_4 = 1$(discussed above)

For $Q_4 = 1$ and $R_4 = 5$

$$\begin{array}{r|l|l} 5^{24} & 7\ {}_2 2\ {}_5 3\ {}_3 8\ {}_5 7 & 61 \\ \hline & 1381 & \end{array}$$

Step 6: Gross dividend = 57

Net dividend = 57 – (sum of cross product of 24 and 81)

= 57 – (2 × 1 + 4 × 8)

= 57 – 34

= 23

Divide 23 by 5

$Q_5 = 4$ and $R_5 = 3$

Remainder = 361 – (sum of cross product of 24 and 14) × 10 – (last flag digit × last quotient)

= 361 – (8 + 4) × 10 – (4 × 4)

= 361 – 120 – 16 = 225

$$\begin{array}{r|l|l} 5^{24} & 7\ {}_2 2\ {}_5 3\ {}_3 8\ {}_5 7 & 61 \\ \hline & 1\ 3\ 8\ 1\ 4 & 225 \end{array}$$

Hence Quotient = 13814 and R = 225

Example: Divide 13579 by 975

Solution: If we take two digits as dhwajanka, then the **remainder** column will certainly have two digits. See the arrangement **below**.

$$9^{75} \mid 1\ 3\ 5 \mid 79$$

Dhwajanka = 75, so last two digits (79) are put in the remainder column

Step 1: Divide 13 by 9

Q_1 = 1 and R_1 = 4

$$9^{75} \mid 1\,3\ {}_4 5 \mid 79$$

Step 2: Gross dividend = 45

Net dividend = 45 – 1 × 7 =38

Divide 38 by 9

Q_2 = 4 and R_2 = 2

9^{75}	1 3 $_4$5	$_2$79
	14	

Step 3: Gross dividend = 27

Net dividend = 27 – (sum of cross product of 15 and 14)

27 – (4 × 7 + 5 × 1)

27 – 33 = – 6

Since the dividend (–6) is negative, so the preceding dividend cannot be taken as 4.

Divide 38 by 9

Hence, **Revised Q_2 = 3 and R_2 = 38 – 9 × 3 = 11**

9^{75}	1 3 $_4$5	$_{11}$79
	13	

Gross dividend in the remainder column =1179

Remainder = 1179 – (3 × 7 + 1 × 5) × 10 -3 × 5

= 1179 – 260 -15

= 904

Quotient = 13 and Remainder = 904

Exercise for Practice

Exercise 4.1

Divide by Nikhilam Method

(a) 4532 ÷ 854 (b) 2101532 ÷ 879

(c) 12345 ÷ 8888 (d) 11203 ÷ 8999

(e) 2100012 ÷ 8997 (f) 2002002 ÷ 89997

Exercise 4.2

Divide by Paravartya Method

(a) 13999 ÷ 112 (b) 11329 ÷1132

(c) 239479 ÷ 11203 (d) 13905 ÷ 113

(e) 39999 ÷ 1812 (f) 5012 ÷ 182

Exercise 4.3

Divide by Dhwajanka Method

(a) 2112 ÷ 87 (b) 13385 ÷ 131

(c) 123456 ÷ 91 (d) 92358 ÷ 151

(e) 87432 ÷ 1192 (f) 468625 ÷ 354

(g) 7932584 ÷ 368 (h) 46823564 ÷ 3542

□□□

5

Casting Out Nines

Introduction

When you solve a sum, whether it is addition, subtraction, multiplication or division, a doubt constantly haunts you. To avoid any mistake you begin to check every step and thus lose precious time. Don't you think that there should be a single method which is equally beneficial in all the basic mathematical operations? The answer to your question is YES, there is a method to check all the four fundamental mathematical operations i.e. addition, subtraction, multiplication and division and that too in a few seconds. This wonder method is called **Casting out Nines** or **Nines-Remainder Method.**

How does this method work?

This is a big question, but you will be surprised to know that this method is so simple that even a primary school-going kid can understand and check his/her calculation.

Casting out Nines literally means to throw nines. Now let us focus on its working.

- Add the digits of a number across, dropping out 9, to get a single figure. If it is not a single figure, add the digits obtained so as to get a single figure between 0 to 8.
- 9 is not taken into account in this process, as a digit sum of 9 is the same as a digit sum of zero.

Before coming to the actual problem, shouldn't we understand the basic meaning of the above two points? Let us take some examples to understand what meaning the above line has.

Example: Find the digit sum of 54653

Verification: Digit sum of 54653= 5 + 4 + 6 + 5 + 3 = 23

Since 23 is a double figure number so to get a single figure, we have to sum it again.

Digit sum of 23 = 2 + 3 = 5

The digit sum of 54653 can be done in other ways very easily. As discussed above, we need not take 9 into account

Digit sum of 54653 = 5 + 4 + 6 + 5 + 3

The two group of number 5 + 4 and 6 + 3 can easily be left out while finding the digit sum of 54653, as their sum is equal to 9.

Example: Find the digit sum of 438219

Verification: Add all the digits

4 + 3 + 8 + 2 + 1 + 9 = 27

Digit sum of 27 = 2 + 7 = 9

As discussed above the digit sum of 9 is equal to that of 0, hence the final digit sum of 438219 is 0.

This can be further simplified:

= 4 + 3 + 8 + 2 + 1 + 9 = 0

Example: Find the digit sum of 4379348568219

Verification: Add all the digits

4 3 7 9 3 4 8 5 6 8 2 1 9

The four groups of numbers have the digit sum 9 which need not be taken into account. Hence the final sum = digits left out

= 4 + 3 + 8 = 15

Digit sum of 15 = 1 + 5 = 6

The above examples are sufficient to understand the basic facts behind the casting out nines method. Now let us take a few examples and check the validity of this rule in each case.

Check for Addition

Whatever we do to the numbers, we also do to their digit sum; then the result that we have from the digit sum of the numbers must be equal to the digit sum of answer.

Example: Verify 47385 + 69384 + 58769+ 38173 + 29464 = 243144

Verification:

	Digit sum of number
47385	0 (Digit sum of 9 is 0)
69384	3
58769	8
38173	4
+ 29464	7
243175	?

LHS = Digit sum of 243175= 2 + 4 + 3 + 1 + 7 + 5 = 22

22 = 2 + 2 = 4

RHS = Sum of the Digit sum of number = 0 + 3 + 8 + 4 + 7 = 22 = 2 + 2 = 4

Since LHS = RHS

Result verified

Example: Verify 87643 + 84397+ 38549 + 29765 = 240354

Verification	Digit sum of number
87643	1
84397	4
38549	2
+ 29765	2
240354	?

LHS = Digit sum of 240354 = 2 + 4 + 0 + 3 + 5 + 4 = 0

RHS = Sum of the Digit sum of number = 1 + 4 + 2 + 2 = 0

Since LHS = RHS

Result verified

Example: Verify 37995 + 68986 + 75438 + 58909 = 241348

Verification:	Digit sum of number
37995	6
68986	1
75438	0
+ 58909	4
241348	?

LHS = Digit sum of 241348 = 2 + 4 + 1 + 3 + 4 + 8 = 4

RHS = Sum of the Digit sum of number = 6 + 1 + 0 + 4 = 2

Since LHS $\neq$ RHS

Hence sum is incorrect

Example: Verify 47385 + 69348 + 58769 + 38173 + 29469 =243548

Verification:	Digit sum of number
47385	0
69348	3
58769	8
38173	4
+ 29469	3
243548	?

LHS = Digit sum of 243548 = 2 + 4 + 3 + 5 + 4 + 8 = 8

RHS = Sum of the Digit sum of number = 0+ 3 + 8 + 4+ 3 = 0

Since LHS $\neq$ RHS

Sum is incorrect

Check for Subtraction

The process of subtraction is the same as applied in addition. Remember that the value of the digit sum of minuend should be greater than that of subtrahend.

Example: Verify the result 8934 – 6758 = 2176

Verification: Digit sum of number

	Digit sum of number
8934	6
– 6758	8
2176	?

Since the digit sum of Minuend is less than the digit sum of Subtrahend, hence we need to replace the digit sum of 8934 in such a way that the final digit sum remains the same.

Here the value of the digit sum of 8934 = 6

This digit sum 6 can also be written in so many ways as the digit sum of 15, 24, 33, 42, 51 and 60 also gives the same value 6. Therefore, the need of the hour is to replace the digit sum of 8934 with any of the given values 15, 24, 33, 42, 51 or 60.

	Digit sum of number
8934	15 or 24 or 33 or 42 or 51 or 60
– 6758	8
2176	?

LHS = Digit sum of 2176 = 2 + 1 + 7+ 6 = 7

RHS = Digit sum of (15 – 8) = 7

Or, Digit sum of (24 – 8 = 16) = 1 + 6 = 7

Or, Digit sum of (33 – 8 = 25) = 2 + 5 = 7

Or, Digit sum of (42 – 8 = 34) = 3 + 4 = 7

Or, Digit sum of (51 – 8 = 43) = 4 + 3 = 7

Or, Digit sum of (60 – 8 = 52) = 5+ 2 = 7

In all the above cases

LHS = RHS

Hence result verified.

Example: Verify 643245 – 368249 = 274996

Verification: Digit Sum of number

	Digit Sum of number
643245	6
– 368249	5
274996	?

LHS = Digit sum of 274996 = 1

RHS = 6 – 5 = 1

Since LHS = RHS

Hence, result verified

Example: Verify 2684352485 – 1296827020 = 273514996

Verification:

	Digit Sum of number
2684352485	2
– 1296827020	1
1735149915	?

LHS = Digit sum of

1735149915 = 1 + 7 + 3 + 5 +1+ 4 + 9 + 9 + 1 + 5 = 0

RHS = 2 – 1 = 1

Since LHS ≠ RHS

Hence, result is incorrect.

Check for Multiplication

Multiplication is the most error-prone fundamental operation in mathematics. Students always have doubts about the accuracy of their result and waste time in re-checking every operation again and again. This method will prove a panacea for all those who are not very much sure about their result. Let us take a few examples to understand the *modus operandi* of this method.

We know,

Multiplicand × Multiplier = Product

Example: Verify 12876 × 43853 = 564651228

Verification:

Digit sum of Multiplicand = 1 + 2 + 8 + 7 + 6 = 6

Digit sum of Multipliers =4 + 3 + 8 + 5 + 3 = 5

LHS = Digit sum of Multiplicand and Multiplier taken together

(6 × 5 = 30) = 3

RHS= Digit sum of Product

=5 + 6 + 4 + 6 + 5+ 1 + 2 + 2 + 8 = 3

Since LHS = RHS

Result verified

Example: Verify 5972 × 4853 = 29882116

Verification:

Digit sum of Multiplicand = 5 + 9 + 7 + 2 = 5

Digit sum of Multipliers = 4 + 8+ 5 + 3 = 2

LHS = Digit sum of Multiplicand and Multiplier taken together

(5 × 2 = 10) = 1

RHS = Digit sum of Product

= 2 + 9 + 8 + 8 + 2 + 1 + 1 + 6 = 1

Since LHS = RHS

Result verified

Example: Verify 12589476 × 43256853 = 54058115267908

Verification:

Digit sum of Multiplicand =1 + 2 + 5 + 8 + 9 + 4+ 7 + 6 = 6

Digit sum of Multipliers =4 + 3 +2 + 5+ 6+ 8 + 5 + 3 = 0

LHS = Digit sum of Multiplicand and Multiplier taken together

(6 × 0 = 0) = 0

RHS = Digit sum of Product

= 5 + 4 + 0+ 5 + 8 + 1 + 1 + 5 + 2 + 6 + 7+ 9+ 0+ 8 = 7

Since LHS ≠ RHS

Result Incorrect

Check for Division

We know

Dividend = Divisor × Quotient + Remainder

The casting out nines method described in the very beginning will suffice to check the division operation effectively. You have only one thing to do –

Find the digit sum of Dividend, Divisor, Quotient and Remainder and put the value of digits sum in LHS and RHS. If the same digit sum is obtained on both the sides, it ultimately tells you that you have performed the right operation. Let us take some examples to understand how effectively this method works for Division.

Example: Verify 876543 ÷ 123, Q = 7126 and R = 45

Verification:

Here Dividend = 876543

Digit sum of Dividend = 8 + 7 + 6 + 5 + 4 + 3 = 6

Divisor = 123

Digit sum of Divisor = 1 + 2 + 3 = 6

Quotient = 7126

Digit sum of Quotient = 7 + 1 + 2 + 6= 7

Remainder = 45

Digit sum of Remainder = 4 + 5 = 0

Putting the digit sum value in the given formula, we get,

LHS = Digit sum of Dividend = 6

RHS = Divisor × Quotient + Remainder

= 6 × 7 + 0 = 42

Digit sum of 42 = 6

Hence LHS = RHS

Result verified

Example: Verify 8765 ÷ 243, Q = 36 and R = 23

Verification:

Here Dividend = 8765

Digit sum of Dividend = 8 + 7 + 6 + 5 =8

Divisor = 243

Digit sum of Divisor = 2 + 4 + 3 = 0

Quotient = 36

Digit sum of quotient = 3 + 6 = 0

Remainder = 23

Digit sum of Remainder = 2+ 3 = 5

Putting the digit sum value in the given formula, we get,

LHS = Digit sum of Dividend = 8

RHS = Divisor × Quotient + Remainder

= 0 × 0 + 5 = 5

Hence LHS ≠ RHS

Result incorrect

The Casting out Nines method is equally applicable to check square, square root, cube and cube root, depending upon your ability to perform the verification effectively. In one sentence, this method is the best method to check the accuracy of your result in a few seconds, but it all depends on your practice and ability to judge the digit sum and omit 9.

Exercise for Practice

Exercise 5

Verify the following result

(a) 112065 + 360085 + 289872 + 156345 = 918367

(b) 4998 + 6789 + 5715 + 4837 + 8976 = 31315

(c) 7534 + 2459 + 1932 + 6547 = 16472

(d) 37467 + 35647 + 285 + 10085 = 82876

(e) 3746735 – 2837546 = 909189

(f) 876542 – 32548 – 698547 = 145447

(g) 658723 + 154639 – 369847 + 367 = 443882

(h) 588 × 512 = 301056

(i) 842 × 858 = 722536

(j) 966 × 973 = 939918

(k) 13579 ÷ 975, Q = 13, R = 904

(l) 7238761 ÷ 524, Q= 13184, R = 225

(m) 11199171 ÷ 99979, Q = 112, R = 1523

(n) 87265 × 32117 = 2802690005

(o) 6471 × 6212 = 40197852

(p) $(207)^2$ = 42849

(q) $(2134)^2$ = 4553856

(r) $(3247)^2$ = 10542169

❑❑❑

6

Square

Introduction

The square of a number is that number, multiplied by itself. The Greeks were the first to form different numerical patterns using pebbles. They arranged the pebbles in the shape of a triangle, square, pyramid, pentagon, and hexagon. Here are a few square arrangements shown.

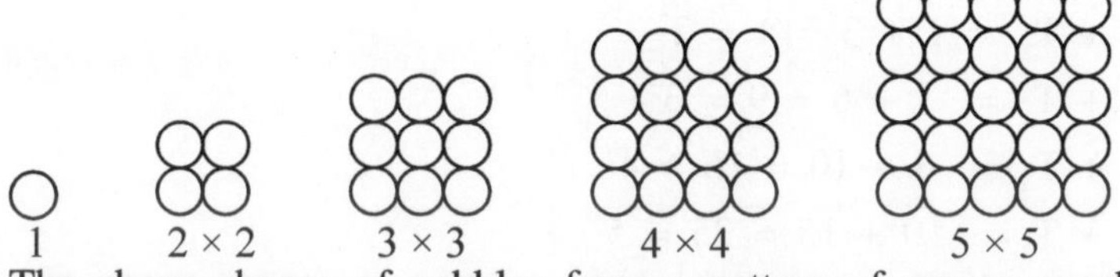

The above shapes of pebbles form a pattern of square numbers. In the language of indices

$$A \times A = A^2$$

$$4 \times 4 = 4^2$$

An Interesting Pattern

$1 = 1 = 1^2$

$1 + 3 = 4 = 2^2$

$1 + 3 + 5 = 9 = 3^2$

$1 + 3 + 5 + 7 = 16 = 4^2$

$1 + 3 + 5 + 7 + 9 = 25 = 5^2$

$1 + 3 + 5 + 7 + 9 + 11 = 36 = 6^2$

$1 + 3 + 5 + 7 + 9 + 11 + 13 = 49 = 7^2$

$1 + 3 + 5 + 7 + 9 + 11 + 13 + 15 = 64 = 8^2$

$1 + 3 + 5 + 7 + 9 + 11 + 13 + 15 + 17 = 81 = 9^2$

$1 + 3 + 5 + 7 + 9 + 11 + 13 + 15 + 17 + 19 = 100 = 10^2$

This pattern shows that the **sum of n odd numbers = n^2**

The arrangement of pebbles in the shape of triangles is called the triangular number.

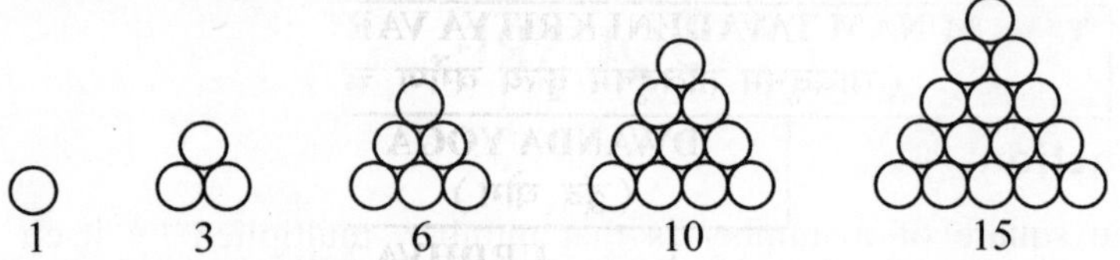

1, 3, 6, 10 and 15, are the first five triangular numbers.

Relation between Square Number and Triangular Number

$T_1 = 1 = 1^2$

$T_1 + T_2 = 1 + 3 = 4 = 2^2$

$T_2 + T_3 = 3 + 6 = 9 = 3^2$

$T_3 + T_4 = 6 + 10 = 16 = 4^2$

$T_4 + T_5 = 10 + 15 = 25 = 5^2$

$T_5 + T_6 = 15 + 21 = 36 = 6^2$

$T_6 + T_7 = 21 + 28 = 49 = 7^2$

$T_7 + T_8 = 28 + 36 = 64 = 8^2$

$T_8 + T_9 = 36 + 45 = 81 = 9^2$

$T_9 + T_{10} = 45 + 55 = 100 = 10^2$

Vedic Method of Squaring

Vedic mathematics has transformed the very technique of squaring. The simple straight multiplication or diagonal method taught in our school will prove insipid, once you learn the Vedic method for squaring. Moreover, there are a number of methods of squaring a number in Vedic mathematics. Depending upon your needs, you can choose the best suited for you.

Vedic Sutras for Squaring

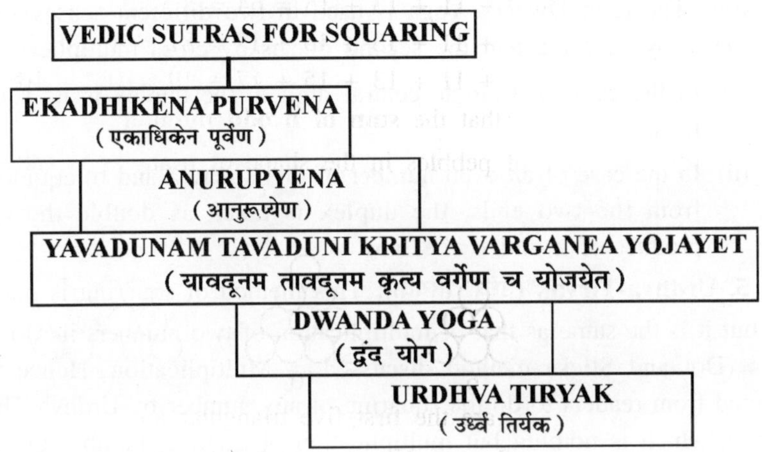

Meaning of Vedic Sutras

1. Ekadhikena Purvena (एकाधिकेन पूर्वेण): The word Ekadhikena Purvena literally means – **one more than the previous one.** This Vedic sutra for squaring is valid only if a number ends with 5.

2. Anurupyena (आनुरूप्येण): This sutra simply means – **"Proportionately"** and is very useful when used for squaring a two digit number. The detailed discussion on its working will be provided in due course.

3. Yavadunam Tavaduni Krtiya Varganaca Yojayet (यावदूनम तावदूनम कृत्य वर्गेण च योजयेत्): This Vedic sutra simply says –

(i) Find the extent or deficiency of a number to be squared with respect to its base. This extent or deficiency is termed here the **deviation.**

(ii) Set up the square of the deviation at the end.

This Vedic sutra works better when the number to be squared is nearer to the base, which is a multiple of 10.

4. Dwanda-Yoga (द्वंद योग) or Duplex Method: The methods discussed so far have limited applicability. The Duplex combination is applicable in all the cases. The term Dwanda-Yoga is used in two different senses – The first one is by squaring and the second one is by cross multiplication.

(i) In the case of a single central digit **a,** the duplex is its square. i.e. $\mathbf{a^2}$

(ii) In the case of an even number of digits (say a and b) equidistant from the two ends, the duplex is taken as double the cross product of a and b (i.e.2ab)

5. Urdhva-Tiryak (उर्ध्व तिर्यक): This method of squaring is nothing new but it is the same as that of multiplication of two numbers in Dot and Cross (Dot and Stick) method discussed in Multiplication. Hence it is expected from readers to do the squaring of any number by Urdhva-Tiryak Method, which is nothing but multiplication of a single number twice.

Applications of Vedic Sutras

1. Ekadhikena Purvena (एकाधिकेन पूर्वेण)

Earlier it was mentioned that this method of squaring is applicable only when the unit digit of any number ends with 5. Let us take some examples to understand its working.

The *modus operandi* of this method is as follows:

- Square the unit digit.
- Multiply the remaining digit by number one more than itself.

Square of a number ending with 5
= Left digit(s) × {Left digit(s) + 1} / 25

A simple example will suffice to clarify the *modus operandi* of this method.

Example: Find the square of 25?

Solution: Since the unit digit is 5, the Ekadhikena sutra will be applicable to this case.

Square of unit digit = 5^2 = 25

Multiply 2 with (2 + 1 = 3) = 2 × 3 = 6

Hence $(2\ 5)^2 = 625$

Example: $(35)^2$ = ?

Solution: Squaring the unit digit = 5^2 = 25

Multiplying 3 with one more than 3 (3+1 = 4) = 3 × 4 = 12

Hence $(35)^2$ = 1225

Look at the following examples:

$(45)^2$ = 4 × 5 | 25 = 2025

$(55)^2$ = 5 × 6 | 25 = 3025

$(65)^2$ = 6 × 7 | 25 = 4225

$(75)^2$ = 7 × 8 | 25 = 5625

$(85)^2$ = 8 × 9 | 25 = 7225

$(95)^2$ = 9 × 10 | 25 = 9025

$(105)^2$ = 10 × 11 | 25 = 11025

2. Anurupyena **(आनुरूप्येण)**

This sub-sutra is helpful in finding the square of a two-digit number. The process of squaring can be summarised in the following steps.

- It consists of three columns.
- The 1st column consists of squaring of digit placed at tens.
- The 2nd column contains the product of unit digit and ten's digit.
- The 3rd column contains the square of unit digit.
- Double the figure written in 2nd column and add from right to left keeping only one digit in 3rd and 2nd column.

If you look upon the working procedure summarised in the above steps, you will notice that it is nothing but the application of algebraic identities – $(a + b)^2 = a^2 + 2ab + b^2$

Example: Find the square of 75 by Anurupyena sutra.

Solution:

COLUMN 1	COLUMN 2	COLUMN 3
$(7)^2 = 49$	$7 \times 5 = 35$	$(5)^2 = 25$
+	$7 \times 5 = 35$	
49	**70**	**25**

= 49 + 7 | 0 + 2 | 5
= 5625

Example: $(87)^2$ = ?

Solution:

COLUMN 1	COLUMN 2	COLUMN 3
$(8)^2 = 64$	$8 \times 7 = 56$	$(7)^2$
+	$8 \times 7 = 56$	
64	**11 2**	**49**

= 64 + 11 | 2 + 4 | 9
= 7569

Example: $(98)^2$ = ?

Solution:

COLUMN 1	COLUMN 2	COLUMN 3
$(9)^2 = 81$	$9 \times 8 = 72$	$(8)^2 = 64$
	$9 \times 8 = 72$	
= 64	**14 4**	**64**

= 81 | 144 + 6 | 4
= 81 | **15** 0 | 4
= 81 + 15 | 0 | 4
= 9604

3. Yavadunam-Tavduni Kritya Varganaca Yojayet
(यावदूनम तावदूनम कृत्य वर्गेण च योजयेत्)

This Vedic sutra literally means – Whatever the excess or deficiency of a number is with respect to its base; exceed or less it further to that extent and set up the square of the deviation at the end. The simplified explanation has been made at the beginning of this chapter. The working of this method has been illustrated in the following examples which are self-explanatory.

Example: Find the square of 12

Solution: The nearest base to 12 is 10.

Deviation from the base = 12 – 10 = 2

Add 2 to 12 and set up the square of 2 at the end.

= 12 + 2 | 2^2

= 144

Example: Find the square of 15

Solution: The nearest base to 15 is 10.

Deviation from the base = 15 – 10 = 5

Add 5 to 15 and set up the square of 5 at the end.

= 15 + 5 | 5^2

= 20 | 25 = 225

(+)

Example: Find the square of 91.

Solution: The nearest base to 91 is 100.

Deviation from the base = 91 – 100 = – 9

Subtract 9 from 91 and set up the square of (–9)at the end.

= 91 – 9 | 9^2

= 82 | 81

= 8281

Example: Find the square of 108

Solution: The nearest base to 108 is 100.

Deviation from the base = 108 – 100 = 8

Add 8 to 108 and set up the square of 8 at the end.

= 108 + 8 | 8^2

= 116 | 64

= 11664

As far as the digit at the right side of the vertical separator is concerned, the rule of Multiplication will follow.

(For more information refer **Table 1** in the chapter on Multiplication.)

Example: Find the square of 997.

Solution: The nearest base to 997 is 1000.

Deviation from the base = 997 – 1000 = -003

Subtract 003 from 91 and set up the square of (-003) at the end.

$= 997 - 003 \mid (-003)^2$

$= 994 \mid 009$

$= 994009$

Though this sutra works better when the number to be squared is nearer to the base, which is in the power of 10(10, 100, 1000….), but in case the base is in the multiple of 10, such as 20(10 × 2), 30(10 × 3), 40(10 × 4) etc, multiply the extreme left figure by 2, 3, 4... (the multiple of base). Let us see a few more examples:

Example: Find the square of 32.

Solution: 32 is no where nearer to 10, hence 30 (= 10 × 3) is taken as the working base

Deviation from the base = 32 – 30 = 2

$(32)^2 = 3 \times (32 + 2) \mid 2^2$

$= 3 \times 34 \mid 4$

$= 102 \mid 4$

$= 1024$

Example: Find the square of 48.

Solution: 48 is no where nearer to 10, hence 50 (= 10 × 5) is taken as the working base

Deviation from the base = 48 – 50 = – 2

Hence $(48)^2 = 5 \times (48 - 2) \mid 2^2$

$= 5 \times 46 \mid 4$

$= 2304$

Example: Find the square of 75.

Solution: 75 is nearer to working base 70 (= 10 × 7)

Deviation from the base = 75 – 70 = 5

$(75)^2 = 7 \times (75 + 5) \mid 5^2$

$= 7 \times 80 \mid 25$

$= 56\mathbf{0} \mid 25$ (2 carried and added to 0)

$= 5625$

4. Duplex or Dwanda Yoga (द्वंद योग) Method

The Dwanda Yoga or Duplex method of squaring is one of the best squaring method in Vedic Mathematics. This is unique in the sense that it has universal application. The most interesting part of this method is that it is based on the Urdhva-Tiryak principle. The group formation in Duplex is same as done in the dot and cross method under Urdhva-Tiryak sutra in multiplication.

Before discussing this new method, we shall deal with the duplex of few numbers and learn the group formation used in Duplex Method.

- Duplex of 1 digit number = Square of that number

  Duplex of 4 = 4^2 = 16

- Duplex of 2-digit number = 2 × (Product of digits)

  Duplex of 24 = 2 × (2 × 4) = 16

- Duplex of 3-digit number

  = 2 × (1st digit × 3rd digit) + (square of middle digit)

  Duplex of 256 = 2 × (2 × 6) + 25 = 49

- Duplex of 4-digit number

  = 2 × (1st digit × 4th digit) + 2 × (2nd digit × 3rd digit)

  Duplex of 5678 = 2 × (5 × 8) + 2 × (6 × 7) = 164

- Duplex of 5-digit number

  = 2 × (1st digit × 5th digit) + 2 × (2nd digit × 4th digit) + $(\text{middle digit})^2$

  Duplex of 56789 = 2 × (5 × 9) + 2 × (6 × 8) + 49 = 235

Once you learn to find the duplex of a number, you need to write the number in groups. The following pattern will help you in grouping the numbers.

$(11)^2 = 1\ 2\ 1$

$(1\ 1\ 1)^2 = 1\ 2\ 3\ 2\ 1$

$(1\ 1\ 1\ 1)^2 = 1\ 2\ 3\ 4\ 3\ 2\ 1$

$(1\ 1\ 1\ 1\ 1)^2 = 1\ 2\ 3\ 4\ 5\ 4\ 3\ 2\ 1$

Grouping of a number

The grouping of $(24)^2$ will follow the pattern of 1 2 1 of $(11)^2$

The groups for **24** are –

$$\underbrace{2}_{\text{1 digit}} \quad \underbrace{24}_{\text{2 digits}} \quad \underbrace{4}_{\text{1 digit}}$$

The grouping of $(245)^2$ will follow the pattern of 1 2 3 2 1 of $(111)^2$.

The groups of numbers for **245** are –

$$\underbrace{2}_{\text{1 digit}} \quad \underbrace{24}_{\text{2 digits}} \quad \underbrace{245}_{\text{3 digits}} \quad \underbrace{45}_{\text{2 digits}} \quad \underbrace{5}_{\text{1 digit}}$$

The grouping of $(2456)^2$ will follow the pattern of 1 2 34 3 2 1 of $(1111)^2$.

The groups of numbers for **2456** are –

$$\underbrace{2}_{\text{1 digit}} \quad \underbrace{24}_{\text{2 digits}} \quad \underbrace{245}_{\text{3 digits}} \quad \underbrace{2456}_{\text{4 digits}} \quad \underbrace{456}_{\text{3 digits}} \quad \underbrace{56}_{\text{2 digits}} \quad \underbrace{6}_{\text{1 digit}}$$

How does Duplex Method work?

- Form the group of numbers to be squared as shown above.
- Write the duplex of each group.
- Once the duplex value for each group is written, add the figures from right to left, as done in multiplication.

Example: Find the square of 24

Solution: The groups for **24** are –

$$\text{Duplex} = \underbrace{2^2}_{} \quad \underbrace{2\times2\times4}_{} \quad \underbrace{4^2}_{}$$

(groups: 2, 24, 4)

= 4 | 16 | 16 (carry +)

= 576

Example: Find the square of 75

Solution: The groups for **75** are

$$\text{Duplex} = 7^2 \quad 2\times7\times5 \quad 5^2$$

(groups: 7, 75, 5)

= 49 | 70 | 25 (carry +)

= **5625**

Example: Find the square of 245.

Solution: The groups of numbers for **245** are –

$$\text{Duplex} = 2^2 \quad 2\times2\times4 \quad 2\times2\times5+4^2 \quad 2\times4\times5 \quad 5^2$$

(groups: 2, 24, 245, 45, 5)

= 4 | 16 | 36 | 40 | 25 (carry +)

= 60025

Example: Find the square of 268

Solution: The groups of numbers for **268** are –

$$\text{Duplex} = 2^2 \quad 2\times2\times6 \quad 2\times2\times8+6^2 \quad 2\times6\times8 \quad 8^2$$

(groups: 2, 26, 268, 68, 8)

= 4 | 24 | 68 | 96 | 64 (carry +)

= 71824

Example: Find the square of 1234

Solution: The groups of numbers for 1234 are –

1, 12, 123, 1234, 234, 34, and 4.

Duplex of 1 = 1^2 = 1

Duplex of 12 = 2 × 1 × 2 = 4
Duplex of 123 = 2 × 1 × 3 + 2^2 = 10
Duplex of 1234 = 2 × 1 × 4 + 2 × 2 × 3 = 20
Duplex of 234 = 2 × 2 × 4 + 3^2 = 25
Duplex of 34 = 2 × 3 × 4 = 24
Duplex of 4 = 4^2 = 16

Arrange the value of duplex as follows-

1 | 4 | 10 | 2 0 | 2 5 | 2 4 | 1 6

Hence $(1234)^2$ = 1522756

Example: Find the square of 23456

Solution: The groups for 23456 are

2, 23, 234, 2345, 23456, 3456, 456, 56 and 6

Duplex of 2 = 2^2 = 4
Duplex of 23 = 2 × 2 × 3 = 12
Duplex of 234 = 2 × 2 × 4 + 3^2 =25
Duplex of 2345 = 2 × 2 × 5 + 2 × 3 × 4 =44
Duplex of 23456 = 2 × 2 × 6 + 2 × 3 × 5 + 4^2 =70
Duplex of 3456 = 2 × 3 × 6 + 2 × 4 × 5 = 76
Duplex of 456 = 2 × 4 × 6 + 5^2 = 73
Duplex of 56 = 2 × 5 × 6 = 60
Duplex of 6 = 6^2 = 36

Arrange the duplex of each number in digit separator.

= 4 | 1 2 | 2 5 | 4 4 | 7 0 | 7 6 | 7 3 | 6 0 | 3 6

= 5 5 0 1 8 3 9 3 6

Proceeding in the same fashion, readers can find the square of any number of their choice by using Duplex Method. All the methods discussed above have their own beauty and it is now for the readers to decide which method is the best according to a particular situation.

EXERCISE FOR PRACTICE

Exercise 6.1

Find the square of the following by Ekadhikena Sutra

(a) 15 (b) 25 (c) 35 (d) 65
(e) 85 (f) 95 (g) 105 (h) 115
(i) 125

Exercise 6.2

Find the square of the following by Anurupyena Sutra

(a) 44 (b) 47 (c) 68 (d) 57
(e) 89 (f) 76 (g) 53 (h) 59
(i) 67

Exercise 6.3

Find the square of the following by Yavadunam Sutra

(a) 15 (b) 18 (c) 35 (d) 48
(e) 54 (f) 58 (g) 92 (h) 112
(i) 107

Exercise 6.4

Find the square of the following by Dwanda Yoga Sutra

(a) 29 (b) 34 (c) 47 (d) 3
(e) 58 (f) 79 (g) 85 (h) 97
(i) 99 (j) 109 (k) 168 (l) 198
(m) 2356 (n) 6254 (o) 2591

❑❑❑

7

Square Root

Introduction

The square root of a number **a** is that number which when multiplied by itself, gives **a** as the product.

In other words, if × is the square root of y then,

$$x^2 = y$$

i.e. $x = \sqrt{y}$ or $(y)^{\frac{1}{2}}$

You must have learnt two methods commonly taught in our secondary school syllabus, to find the square root of any number.

They are –

(a) Factor Method

(b) Division Method

Both these methods are time-consuming and boring. Vedic Mathematics provides several Vedic Sutras to find the square root of any given number. Some of these Vedic Sutra which we shall discuss in this chapter are not only easy to understand, but save our precious time and energy. Besides Vedic Sutras to find the square root of any number, we shall also learn Aryabhatta's method of finding square root, which is quite similar to Dwanda-yoga (**द्वंद योग**) Vedic sutra.

Vedic Sutra for Square Root

The best method taught in our school syllabus for finding the square root is Long Division Method. Here we shall learn the best Vedic Method called **Dwanda-Yoga (द्वंद योग) or Duplex Method** to find the square root of a number just by small mathematical calculation.

Meaning of Vedic Sutra

The literal meaning of Dwanda-Yoga (द्वंद योग) has been discussed in Chapter-6 (Square), please refer to it for more information.

Some Important Points to Remember:

- If the unit digit of any number is 2, 3, 7, or 8, it can't be a perfect square root.
- If a number ends in an odd number of zeros, it can't be a perfect square root.

Unit digit of a square root	Unit digit of a square
0	0
1	1 or 9
4	2 or 8
5	5
6	4 or 6
9	3 or 7

The above table shows:

- A complete square ending in 1 must have either 1 or 9, mutual complements from 10, as the last digit of its square root.
- A square can end in 4, only if the square root ends in 2 or 8, a complement from 10.
- If a square ends with 5, its square root too ends with 5.
- A square ending in 6 must have 4 or 6, a complement of 10, as the last digit in its square root.

- The termination of an exact square in 9 is possible, if and only if square root ends in 3 or 7.

Aryabhatta's Method of Finding Square Root

भागं हरदेवर्गान्नित्यं द्विगुणेन वर्गमूलेन ।

वर्गा द्वये शुद्धे लब्धं स्थानान्तरे मूलम।।

This is the fourth sloka of Ganita Pada, written by the great Indian astronomer and mathematician Aryabhatta born in 476 AD. The above mentioned fourth sloka describes the method of finding the square root and consists of the following steps:

- Subtract the greatest possible square from the last odd place.
- Always divide the even place by twice the square root of the preceding odd place.
- Subtract from the odd place (standing on the right) the square of the quotient.
- Repeat the process as long as there are still digits on the right.

This method is illustrated with the help of some examples which are self-explanatory in nature.

Example: Find the square root of 529.

Solution:

- First form the group of numbers. Mark the even places by horizontal lines and odd places by vertical lines.

| _ |

5 2 9

- Since $4 < 5 < 9$ or $2 < \sqrt{5} < 3$, so that the integral part of $\sqrt{5}$ is 2.
- Subtracting the square of the root.

$$\begin{array}{r|l} \overset{|}{5}\ 2\ \overset{|}{9} & \underline{2} \\ -\underline{4} & \end{array}$$

Divide by $2 \times 2 = 4 \overline{\smash{\big)}\, 1\ 2} \, 3$

$\quad - 1\ 2$

$\quad\quad 9$

Subtracting the square of quotient. -3^2

$\quad \times$

Subtracting the square of quotient.

As no more places are left on the right, the process ends.

The required square root of 529 is 23.

Example: Find the square root of 1225.

Solution:

- First form the group of numbers. Mark the even places by horizontal lines and odd places by vertical lines.

  $1\,\overset{|}{2}\,\bar{2}\,\overset{|}{5}$

- Since $9 < 12 < 16$ or $3 < \sqrt{12} < 4$, so the integral part of $\sqrt{12}$ is 3.
- Subtracting the square of the root

$1\,\overset{|}{2}\,\bar{2}\,\overset{|}{5} \,|\, 3$

-9

Divide by $2 \times 3 = 6 \overline{\smash{\big)}\, 3\ 2} \quad 5$

$-3\ 0$

Subtracting the square of quotient $\quad 2\ 5$

-5^2

$\times$

Hence the square root of 1225 is 35

Example: Find the square root of 119716.

Solution:

- First form the group of numbers. Mark the even places by horizontal lines and odd places by vertical lines.

  $\overset{|}{11}\ \overset{-}{9}\ \overset{|}{7}\ \overset{-}{1}\ \overset{|}{6}$

- Since 9 < 11 < 16 or $3 < \sqrt{11} < 4$, so the integral part of "$\sqrt{11}$ is 3.

Subtracting the square of the root

$$\begin{array}{l} |\ -\ |\ -\ | \\ 11\ 9\ 7\ 1\ 6 \mid 3 \\ -9 \end{array}$$

Divide by $2 \times 3 = 6$ | 2 9 ; $-2\ 4$ | 4 ; 5 7

Subtracting the square of quotient -4^2 ; 41

Divide by $2 \times 34 = 68$ | 4 1 1 | 6 ; $-\ 4\ 0\ 8$; 3 6

Subtracting the square of quotient $-\ 6^2$; X

As no more places are left on the right, the process ends.

The required square root of 119716 is 346.

Example: Find the square root of 738915489

Solution: Placing the horizontal lines on even places and vertical lines on the odd places, we have

| – | – | – | – |
7 3 8 9 1 5 4 8 9

The integral part of $\sqrt{7}$ is 2 as $2^2 < 7 < 3^2$.

Subtract the square of the integral part

| 7 3 8 9 1 5 4 8 9 | 2
-4

Divide by $2 \times 2 = 4$ | 3 3 | 7 ; $-\ 2\ 8$; 5 8

Subtract the square of the quotient $-\ 7^2$

Divide by $2 \times 27 = 34$ | 9 9 | 1
− 5 4
4 5 1

Subtract the square of the quotient − 1^2

Divide by $2 \times 271 = 542$ | 4 5 0 5 | 8
− 4 3 3 6
1 6 9 4

Subtract the square of the quotient − 8^2

Divide by $2 \times 2718 = 5436$ | 1 6 3 0 8 | 3
− 1 6 3 0 8
9

Subtract the square of the quotient − 3^2
X

Hence the required square root = 7183

Important Point to Note : When divided by 4 we avoid to take 8 as quotient, as at the next stage we will get $18 - 8^2 = -46$, a negative number.

Vedic Method for Square Root

The best and simplest Vedic Method for finding the square root is **Dwanda-Yoga or Duplex Method**.

Fundamental rules for extraction of square root of any number

- The given number is first arranged in two-digit groups from right to left and a single digit if any left over at the left hand end is counted as a simple group by itself.
- The number of digits in the square root will be the same as the number of digit-groups in the given number itself including a single digit if any such there be.
- If a square root contains n digits, the square must consist of 2n or 2n − 1 digits.

- And conversely, if the given number has n digits, the square root will contain n| 2 or (n + 1)|2 digits.

Working Procedure of Duplex Method

The working of Duplex method is as simple as the straight division. The Vedic Method of finding the square root is a treatment over Aryabhatta's method of extracting the square root.

Let us take a few examples to understand the *modus operandi* of this method.

Example: Find the square root of 529 by Duplex method.

Solution:

- Group the number 5 2 $\overline{9}$ by placing a bar over it.
- Put horizontal and vertical lines as shown below.

	5	29
Q		

- Since $2^2 < 5 < 3^2$, so the first digit of square root in quotient column is 2. Double the quotient and set down as divisor

4	5	29
Q	2	

Remainder = $5 - 2^2 = 1$

Put this remainder below the next dividend digit. Hence the next gross dividend = 12

4	5	$_1$29
Q	2	3

- Divide 12 by 4 and put the quotient 3 in quotient column.

  $12 \div 4 = 3$

- Next dividend = 9

Subtract the square of quotient from the next dividend.

Net dividend = $9 - 3^2 = 0$

Since no more digits are left, so the square root of 529 is 23.

Example: Find the square root of 1225.

Solution:

- Group the number $\overline{12}\ \overline{25}$, by placing a bar over it.
- Put horizontal and vertical lines as shown below.

	12	25
Q		

- Since $3^2 < 12 < 4^2$, so the first digit of square root in the quotient column is 3. Double the quotient and set down as divisor.

6	12	25
Q	3	

  Remainder = $12 - 3^2 = 3$

  Put this remainder below the next dividend digit. Hence the next gross dividend = 32.

6	12	$_32$5
Q	3	

- Divide 32 by 6 and put the quotient 5 in the quotient column and the remainder 2 below 5.

6	12	$_32\ _25$
Q	3	5

- Next dividend = 25

  Subtract the square of quotient from the next dividend.

  Net dividend = $25 - 5^2 = 0$

  Since no more digits are left, so the square root of 1225 is 35.

Example: Find the square root of 20736

Solution:

- Group the number $\overline{2}\ \overline{07}\ \overline{36}$, by placing a bar over it.
- Put horizontal and vertical lines as shown below.

	2	07 36
Q		

- Since $1^2 \leq 2 < 2^2$, so the first digit of square root in quotient column is 1. Double the quotient and set down as divisor.

2	2	07 36
Q	1	

Remainder = $2 - 1^2 = 1$.

Put this remainder below the next dividend digit. Hence the next gross dividend = 10

2	2	$_{1}07$ 36
Q	1	

- Divide 10 by 2

$10 \div 2 = 5$, Q = 5 and R = 0

Since remainder cannot be taken as 0 until the whole operation completes, hence we need to take the quotient less than 5. This is because we cannot consider the remainder zero in the middle of division.

Hence the **Revised Quotient = 4** and **Revised Remainder = 2**

2	2	$_{1}0\ _{2}7$ 36
Q	1	4

- Next dividend = 27

Subtract the square of quotient from the next dividend.

Net dividend = $27 - 4^2 = 11$

Divide 11 by 2 and write the quotient (Q = 5) and remainder (R = 1) at their respective place.

2	2	$_{1}0\ _{2}7\ \ _{1}36$
Q	1	4 5

- Gross dividend = 13

Net dividend = 13 – (Duplex of 45)

$= 13 - 2 \times 4 \times 5$

$= 13 - 40 = -27 < 0$

(Since Net dividend is less than zero, hence we can't take the quotient (Q = 5) as taken above. Now for 11÷ 2, **Revised Q = 4** and **Revised R = 3**.)

2	2	${}_{1}0\ {}_{2}7\ {}_{3}3\ 6$
Q	1	4 4

- Gross dividend = 33

  Net dividend = 33 – (Duplex of 44)

  $= 33 - 2 \times 4 \times 4$

  $= 33 - 32$

  $= 1$

  Divide **1** by 2 and put the quotient **Q = 0** and remainder **R = 1** in its proper place.

2	2	${}_{1}0\ {}_{2}7\ {}_{1}3\ {}_{1}6$
Q	**1**	**4 4** . 0

- Next gross dividend = 16

  Net dividend = 16 – Duplex of 4

  $= 16 - 4^2 = 0$

  Since no more digits are left, so the square root of 20736 is 144

Example: Find the square root of 119716 by Vedic method.

Solution:

- Group the number $\overline{11}\ \overline{97}\ \overline{16}$, by placing a bar over it.
- Put horizontal and vertical line as shown below.

	11	97 16
Q		

- Since $3^2 < 11 < 4^2$, so the first digit of square root in quotient column is 3. Double the quotient and set it down as divisor.

6	11	97 16
Q	**3**	

Remainder $= 11 - 3^2 = 2$

Put this remainder below the next dividend digit. Hence the next gross dividend = 29

$$\begin{array}{c|c|l|} 6 & 11 & {}_{2}97\quad 16 \\ \hline Q & 3 & \end{array}$$

- Divide 29 by 6 and put the quotient 4 in the quotient column and remainder below 7.

$$\begin{array}{c|c|l|} 6 & 11 & {}_{2}9{}_{5}7\quad 16 \\ \hline Q & 3 & \quad 4 \end{array}$$

- Next dividend = 57

  Subtract the square of quotient from the next dividend.

  Net dividend = 57 – 4^2 = 57 – 16 = 41

  Divide 41 by 6 and write the quotient (6) and remainder (5) at their respective place.

$$\begin{array}{c|c|l|} 6 & 11 & {}_{2}9\,{}_{5}7\ {}_{5}1\,6 \\ \hline Q & 3 & \quad 4\,6 \end{array}$$

- Gross dividend = 51

  Net dividend = 51 – (Duplex of 46)

  = 51 – 2 × 4 × 6 = 3

  Divide it by 6 and put quotient (0) and remainder (3) in its proper place.

$$\begin{array}{c|c|l|} 6 & 11 & {}_{2}9\,{}_{5}7\,{}_{5}1{}_{3}6 \\ \hline Q & 3 & \quad 4\,6.0 \end{array}$$

- Next gross dividend = 36

  Net dividend = 36 – duplex of 6 = 36 – 6^2 = 0

  Since no more digits are left, so the square root is 346

Example: Find the square root of 69906321

Solution:

- Put horizontal and vertical lines as shown below.

$$\begin{array}{c|c|l|} & 69 & 90\ \ 63\ \ 21 \\ \hline Q & & \end{array}$$

- Since $8^2 < 69 < 9^2$

  So first digit of square root = 8

  Divisor = 2 × 8 = 16 Remainder = $69 - 8^2 = 5$

  Next gross dividend = 59

16	69	90 63 21
Q	8	

  Divide 59 by 16 and set quotient and remainder in their respective columns. The Q = 3 and R = 11

16	69	${}_59\ {}_{11}0\ 6\ 3\ 2\ 1$
Q	8	3

- Next gross dividend = 110

  Net dividend = $110 - 3^2 = 110 - 9 = 101$

  Divide 101 by 16 and place Q = 6 and R = 5 at proper place.

16	69	${}_59\ {}_{11}0\ {}_56\ 3\ 2\ 1$
Q	8	3 6

- Next gross dividend = 56

  Net dividend = 56 – Duplex of 36

  = 56 – 2 × 3 × 6 = 56 – 36 = 20

  Divide 20 by 16 and write Q = 1 and R = 4 at proper place.

16	69	${}_59\ {}_{11}0\ {}_56\ {}_43\ 2\ 1$
Q	8	3 6 1

- Next dividend = 43

  Net dividend = 43 – Duplex of 361

  = $43 - (2 \times 3 \times 1 + 6^2)$

  = 43 – 42 = 1

- Divide 1 by 16

  Q = 0 and R = 1

  Next dividend = 12

  Net dividend = 12 – Duplex of 61

  = 12 – 2 × 6 × 1 = 0

Divide 0 by 16

Q = 0 R = 0

16	69	$_5 9\ _{11}0\ _5 6\ _4 3\ _1 2\ _0 1$
Q	8	361.000

Since no more digits are left so the square root is 8361.

Example: Find the square root of 389376

Solution:

12	38	$_2 9\ _5 3\ _1 7\ _1 6$
Q	6	24.00

The one line Vedic solution has the following steps:

- Since there are three groups in the above square root, hence the square root consists of 3 digits.
- First digit = 6
- Remainder = 38 – 6^2 = 2. Write 2 below 9.
- New dividend = 29 and Divisor = 2 × 6 = 12
- For 29 ÷ 12 , Q = 2 and R = 5, write quotient(Q) and remainder(R) at the proper place.
- New dividend = 53 and Corrected dividend = 53 – 2^2 = 49
- For 49 ÷ 12 , Q = 4 and R = 1, write Q and R at its proper place.
- So far we have got three digits in the quotient column, hence the next digit in the quotient column will be the digit placed after decimal.
- New dividend = 17 and remainder = 17 – 2 × 4 x 2 = 1
- New dividend = 16 and the last remainder = 16 – 4^2 = 0
- Square Root = 624

Example: Find the square root of 41254929.

Solution:

12	41	$_5 2\ _4 5\ _5 4\ _2 9\ _1 2\ _0 9$
Q	6	423.000

The step by step illustration is done here for your convenience.

- Number of digits in the square root = 4
- The first digit of square root = 6
- Remainder = 41 – 6^2 = 5, place it below 2.
- Next dividend = 52 and divisor = 2 × 6 = 12
- For 52 ÷ 12 , Q = 4 and R = 4, place it at proper place.
- Next dividend = 45 and corrected dividend = 45 – 4^2 = 29
- For 29 ÷ 12 , Q = 2 and R = 5
- Next dividend = 54 and corrected dividend = 54 – 2 × 2 × 4 = 38
- For 38 ÷ 12 , Q = 3 and R = 2
- Since we have so far got 4 digits in the quotient column, hence the perfect square root is obtained. The next digit in the quotient column will give the remainder, if any.
- Remainder = 29 – 4 × 3 × 2 + 2^2 = 1
- Next dividend = 12 and remainder = 12 – 2 × 3 × 2 = 0
- Next dividend = 9 and the last remainder = 9 – 3^2 = 0

Hence square root = 6423

Square Root of Irrational Number

Many a times, while solving the problems of Mensuration, you must have encountered the need to find the square root of irrational number. This method presented below will help you to find the square root of any irrational number in a few moments, and once you are in running practice, you can find the square root of such irrational number mentally.

Square Root of Irrational Number =

$$\sqrt{\text{Nearest Perfect square}} + \frac{\text{Deviation from nearest Perfect square}}{2 \times \sqrt{\text{Nearest Perfect square}}}$$

Example: Find the square root of 47.

Solution: Perfect square approaching 47 is 49.

Deviation = 47 – 49 = – 2

$$\sqrt{47} = \sqrt{49} - \frac{2}{2\times\sqrt{49}}$$

$$= 7 - \frac{1}{7} = 7 - 0.1428 = 6.8572$$

Example: Find the square root of 174.

Solution: Perfect square approaching 174 is 169

Deviation = 174 – 169 = 5

$$\sqrt{174} = \sqrt{169} + \frac{5}{2\times\sqrt{169}}$$

$$= 13 + \frac{5}{26} = 13.192$$

Example: Find the square root of 147.

Solution: Perfect square approaching 147 is 144.

Deviation = 147 – 144 = 3

$$\sqrt{147} = \sqrt{144} + \frac{3}{2\times\sqrt{144}}$$

$$= 12 + \frac{1}{8} = 12.125$$

Example: Find the square root of 187.

Solution: Perfect square approaching 187 is 196.

Deviation = 187 –196 = – 9

$$\sqrt{187} = \sqrt{196} - \frac{9}{2\times\sqrt{196}}$$

$$= 14 - \frac{9}{28} = 14 - 0.32$$

$$= 13.67$$

Exercise for Practice

Exercise 7.1

Find the Square Root of the following by Aryabhatta's Method.

(a) 4473225 (b) 10329796

(c) 14047504 (d) 2116

(e) 4225 (f) 6889

(g) 59049 (h) 125316

(i) 169744 (j) 1265625

Exercise 7.2

Find the Square Root by Duplex Method.

(a) 99920016 (b) 931225

(c) 45369 (d) 790321

(e) 78978769 (f) 45319824

(g) 3249 (h) 552049

(i) 69906321 (j) 41254929

Exercise 7.3

Find the Square Root of following Irrationals.

(a) 37 (b) 164

(c) 175 (d) 43

(e) 231 (f) 198

(g) 69 (h) 287

(i) 631

❑❑❑

8

CUBE

Introduction

When a number is multiplied by itself three times, the number so obtained is called the cube of that number.

$1 \times 1 \times 1 = 1^3 = 1$

$2 \times 2 \times 2 = 2^3 = 8$

$3 \times 3 \times 3 = 3^3 = 27$

$4 \times 4 \times 4 = 4^3 = 64$

$5 \times 5 \times 5 = 5^3 = 125$

$6 \times 6 \times 6 = 6^3 = 216$

$7 \times 7 \times 7 = 7^3 = 343$

$8 \times 8 \times 8 = 8^3 = 512$

$9 \times 9 \times 9 = 9^3 = 729$

In general, $\mathbf{a \times a \times a = a^3}$

Cubes of larger numbers are rarely used in our daily calculations. We sometimes need the cube value of two-digit numbers and generally multiply the number three times. Vedic Mathematics presents several very interesting methods to find the cubes of any given number in a few seconds. All the methods discussed here are easy to understand and easy to practice.

Characteristics of Cubic Number

1. To find the sum of cubes of n natural numbers, apply this formula.

$$1^3 + 2^3 + 3^3 + ...n^3 = \frac{n^2(n+1)^2}{4}$$

2. Cubic numbers, like the square numbers, have a special relation to the odd number. Try to understand the relation between Cubic Number and Odd Number in the following example –

$1^3 = 1$	***First odd number***
$2^3 = 8 = 3 + 5$	***Sum of next two odd numbers***
$3^3 = 27 = 7 + 9 + 11$	***Sum of next three odd numbers***
$4^3 = 64 = 13 + 15 + 17 + 19$	***Sum of next four odd numbers***
$5^3 = 125 = 21 + 23 + 25 + 27 + 29$	***Sum of next five odd numbers***

3. Between 1 and 100 there are only two Cubic numbers 1 and 64, that are also square numbers.

4. Cubic numbers exhibit very special relation with Triangular numbers.(Triangular numbers are those Polygonal numbers, which when represented in dots give the shape of Triangles. 1,3,6,10,15… are examples of Triangular numbers.)

   If $C_{1,}$ $C_{2,}$ C_3 ….are the first, second, third… Cubic numbers and T_1, T_2, T_3— are the first, second and third Triangular numbers, then they exhibit a unique property:-

$$C_1 = 1 = (T_1)^2$$
$$C_1 + C_2 = 1 + 8 = (T_2)^2$$
$$C_1 + C_2 + C_3 = 1 + 8 + 27 = 36 = (T_3)^2$$
$$C_1 + C_2 + C_3 + C_4 = 1 + 8 + 27 + 64 = 100 = (T_4)^2$$

Algebraic Method to find Cube

In school we learn to find the cube of any number using –

$$(a + b)^3 = a^3 + 3a^2b + 3ab^2 + b^3$$

or, $$(a - b)^3 = a^3 - 3a^2b + 3ab^2 - b^3$$

Let me take one example to show how does this formula works?

Example: Find the cube of 98.

Solution: $(98)^3 = (90 + 8)^3$

$= (90)^3 + 3 \times (90)^2 \times 8 + 3 \times 90 \times (8)^2 + (8)^3$

$= 729000 + 194400 + 17280 + 512$

$= 941192$

You can apply another algebraic identity to find the cube of 98.

$(98)^3 = (100 - 2)^3 = (100)^3 - 3 \times (100)^2 \times 2 + 3 \times 100 \times (2)^2 - (2)^3$

$= 1000000 - 60000 + 1200 - 8 = 941192$

Vedic Sutras for Cube

Unlike traditional method taught in school, Vedic Mathematics has several sutras which are very easy and helpful in finding the cube of any number. Some of the Vedic Sutras discussed in this chapter are as follows:

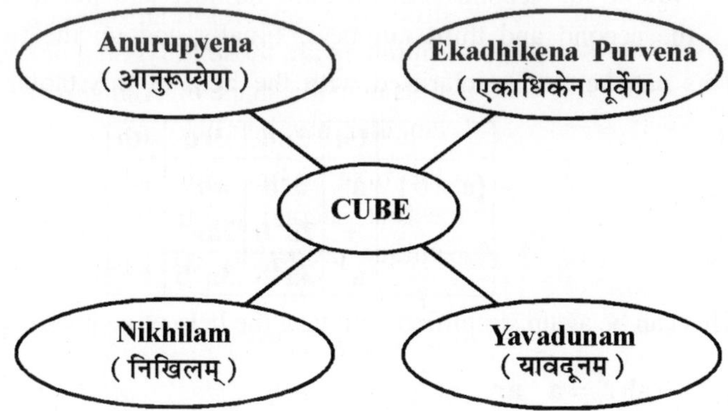

Though all the above sutras have been discussed in the previous chapters, a brief discussion whenever required will be presented.

Anurupyena (आनुरूप्येण) Sutra

The simple meaning of this sutra has been discussed earlier, so I shall throw light only on its working. But before discussing how Vedic Sutra Anurupyena works, let us first understand the concept of **Geometric Progression.** Here are the few examples of geometric series.

(i) 1, 2, 4, 8 ...

(ii) . 2, 8, 32, 128...

In the first example (i):

The first term (a) = 1

$$\text{Common ratio (r)} = \frac{\text{Second Term}}{\text{First Term}} = \frac{\text{Third term}}{\text{Second term}} = ...\frac{2}{1} = \frac{4}{2}... = 2$$

The n th term (T_{n}) of a geometric series = $a\ r^{n-1}$

Where a = first term and r = common ratio.

$T_n = 1\ (2)^{n-1} = 2^{n-1}$

Working of Anurupyena Method

- First take the cube of first digit and multiply it with the common ratio, in a row of 4 figures.
- Double the second and the third number and put it down, under the second and third numbers. Finally add up the two rows.

This can be further clarified with the help of this table.

	1st	2nd	3rd	4th
$(a+b)^3$	a^3	a^2b	ab^2	b^3
	+	$2a^2b$	$2ab^2$	
	a^3	$3a^2b$	$3a^2b$	b^3

This can be again simplified with take the help of geometric progression.

$$\begin{array}{l} (ab)^3 = a \quad ar \quad ar^2 \quad ar^3 \\ \qquad\quad + \ 2ar \quad 2ar^2 \\ \hline \end{array}$$

(Where r = b/a)

N.B. – Here a and b symbolise the two digit number, where a is at ten's place and b is at the unit place.

Let us take a few examples to understand the basic *modus operandi.*

Example: Find the cube of 12

Solution: $(12)^3 = ?$

Here say **a = 1** and **b = 2**

r = common ratio = b/a = 2/ 1 = 2

Hence the table arrangement will be as follows:

$$
\begin{array}{llcccc}
(12)^3 & = & 1 & 2 & 4 & 8 \\
 & & & & 4 & 8 \\
\cline{3-6}
 & & 1 & 7 & 2 & 8
\end{array}
$$

Important Points:

1. *If you start with the cube of first digit and multiply with the geometric ratio up to the next three numbers, the 4th number of the series will be the cube of the 2nd digit.*
2. *Addition should be done from right to left, keeping only single digit at a time and the rest digit will be carried to the next column, and so on.*

Example: Find the cube of 15

Solution: $(15)^3 = ?$

Here say **a = 1** and **b = 5**

r = common ratio = b/a = 5 / 1 = 5

Hence the table arrangement will be as follows –

$$
\begin{array}{llcccc}
(15)^3 & = & 1 & 5 & 25 & 125 \\
 & & & 10 & 50 & \\
\cline{3-6}
 & & 1 & \underline{1}5 & \underline{7}5 & \underline{\mathbf{12}}5 \\
 & & \curvearrowleft\;+ & \curvearrowleft\;+ & \curvearrowleft\;+ &
\end{array}
$$

= 3 3 75

The excess digit, leaving the unit digit from each column (from right to left), is transferred to the next column. In the above example, the excess digit is underlined.

Example: Find the cube of 19

Solution: $(19)^3 = ?$

Here say **a = 1** and **b = 9**

r = common ratio = b/a = 9/ 1 = 9

Hence the table arrangement will be as follows:

```
1     9      81      729
      18     162
--------------------------
1    27     243     729
```

(with arrows marked "+" carrying between columns)

= 6 8 5 9

Example: Find the cube of 32

Solution: $(32)^3 = ?$

Here say **a = 3** and **b = 2**

r = common ratio = b/a = 2/3

Hence the first line of the table arrangement will be as follows:

$(32)^3 = 3^3$ (=27)   27 × 2 / 3 (=18)   18 × 2 / 3 (=12)   12 × 2 / 3 (=8)

```
= 27    18     12     8
        36     24
-------------------------
  27    54     36     8
```

(with arrows marked "+" carrying between columns)

= 3 2 7 6 8

Example: Find the cube of 98

Solution: $(98)^3 = ?$

Here say **a = 9** and **b = 8**

r = common ratio = b/a = 8 / 9

Hence the table arrangement will be as follows:

$(98)^3 = 9^3\,(=729)\quad 729\times 8/9\,(=648)\quad 648\times 8/9\,(=576)\quad 576\times 8/9\,(=512)$

	729	648	576	512
		1296	1152	
=	729	19 44	172 8	51 2
=	729	1944	177 9	2
=	729	212 1	9	2
=	941	192		

The last example has been simplified in all respects. Every step is shown separately for making the concept extra simple to understand. Once you practice a few examples and understand it well, you need not write all the steps and regular practice can save plenty of your time in calculation.

Nikhilam (निखिलम्) Sutra

The *modus operandi* of Nikhilam Sutra can be described in the following steps:

- First take the deviation of the number to be cubed from its base.
- The whole cubing process then involves 3 steps.
  - A) (Number to be cubed + 2 × deviation from the base) × (sub-base)^2
  - B) {3 × (deviation)^2 } × sub-base
  - C) (Deviation)^3
- If there is no sub-base, then the calculation becomes very easy.

Let us take some examples to understand the *modus operandi* of this method.

Example: Find the cube of 15 by Nikhilam Sutra.

Solution: 15 is nearer to the base 10

Deviation = 15 – 10 = 5

$$(15)^3 = \underbrace{15+2\times5}_{\text{1st term}} \mid \underbrace{3\times(5)^2}_{\text{2nd term}} \mid \underbrace{(5)^3}_{\text{3rd term}}$$

$$= 25 \mid 75 \mid 125$$

= 3375

Example: Find the cube of 98 by Nikhilam Sutra

Solution: 98 is nearer to the base 100

Deviation = 98 – 100 = – 2

$$\text{Hence } (98)^3 = \underbrace{98+2\times(-2)}_{\text{1st term}} \mid \underbrace{3\times(-2)^2}_{\text{2nd term}} \mid \underbrace{(-2)^3}_{\text{3rd term}}$$

= 94 | 12 | – 8

= 94 | 11 | 100 – 8

= 94 | 11 | 92

Hence $(98)^3$ = 941192

Example: Find the cube of 109 by Nikhilam Sutra?

Solution: Working base = 100

Deviation = 109 – 100 = 9

$(109)^3$ = 109 + 9 × 2 | 3 × 9^2 | 9^3

= 127 | 2 43 | 7 29

= 1295029

(Since the base =100, so there should be 2 digits in each digit separator)

Important Points

The second and third term should be in accordance to the base, i.e. if the base is 10, the number of digit to be kept in the second and third digit separator should be 1. If the base is 100, the maximum number of digits to be kept in the 2nd and 3rd digit separator should be 2.

Example: Find the cube of 26 by Nikhilam Sutra.

Solution: Since 26 is far from the working base 10, hence we will have to take 20 as a working base.

For $26 = 2 \times 10 + 6$

Base = 10

Sub-base = 2

Deviation = 6

$$(26)^3 = \underbrace{2^2 \times \{26 + 2 \times 6\}} \mid \underbrace{2 \times 3 \times 6^2} \mid \underbrace{6^3}$$

= 152 | **21** 6 | **21** 6

= 17576

Example: Find the cube of 305 by Nikhilam Sutra.

Solution: $305 = 3 \times 100 + 5$

Base = 100

Sub-Base = 3

Deviation = 5

$(305)^3 = (305 + 2 \times 5) \times 3^2 \mid 3 \times 3 \times 5^2 \mid 5^3$

= **2835** | **2** 25 | **1** 25

= 28372625

Ekadhikena Purvena (एकाधिकेन पूर्वेण)

The *modus operandi* of this method is described as follows:

1st part from left = Square of ten's digit × Ekadhika of ten's digit

2nd part from left = Square of ten's digit × Deviation

3rd part from left = 3 × ten's digit × square of unit digit

4th part from left = cube of unit digit

(Here Deviation = 3 × unit digit – 10)

Let us take a few examples to understand the whole procedure.

Example: Find the cube of 43 by Ekadhikena Purvena Sutra

Solution:

	1st part	2nd part	3rd part	4th part
$(43)^3$ =	$4^2 \times (4+1)$	$4^2 \times (3 \times 3 - 10)$	$3 \times 4 \times 3^2$	3^3
	80	–16	108	27
	78*	20 – 16	108	27
	78	4	10 **8**	**2** 7

+ +

$(43)^3 = 79507$

(*2 in the first part = 20 in the second part. Here 80 = 78 + 2 is written)

Example: Find the cube of 78 by Ekadhikena Purvena Sutra

Solution: The whole process of cubing can be written in a tabular form for better understanding.

	1st part	2nd part	3rd part	4th part
$(78)^3$ =	$7^2 \times 8$	$7^2 \times (3 \times 8 - 10)$	$3 \times 7 \times 8^2$	8^3
	392	686	1344	$\underline{51}$ **2**

+

= 392 | 686 | 1344+ 51 | **2**

= 392 | 686 | **139** 5 | **2**

+

= 392 | **82** 5 | 5 | 2

+

= 474552

Example: Find the cube of 93 by Ekadhikena Purvena Sutra

Solution:

	1st part	2nd part	3rd part	4th part
$(93)^3$ =	$9^2 \times (9+1)$	$9^2 \times (3 \times 3 - 10)$	$3 \times 9 \times 3^2$	3^3
	810	-81	243	27
	800	100-81	243	27
	800	19	24 3	**2** 7

+ +

$(93)^3 = 804357$

Yavadunam (यावदूनम) Sutra

The Yavadunam Sutra is also applicable in cubing a number. The only difference here is that we take here not the deficit or the surplus as we do in squaring, but exactly twice the deficit or surplus as the case may be. The whole process can be divided into three parts.

	1st part	2nd part	3rd part
Cube of a number	Number + (2 × excess / deficit from the base)	New excess/deficit × original excess / deficit	Cube of original excess / deficit

The following example illustrates the whole process.

Example: Find the cube of 104 by Yavadunam Sutra

Solution:

The nearest base for 104 is 100.

104 = 100 + 4

Excess from the base = 4

1st part = Number + 2 × Excess

= 104 + 2 × 4

= 112

2nd part = New excess × Original excess

= (112 – 100) × 4

= 12 × 4

= 48

3rd part = Cube of excess

= 4^3

= 64

Arrange the three parts as follows

112 | 48 | 64

Hence $(104)^3$ = 1124864

Example: Find the cube of 113 by Yavadunam Sutra

Solution:

The nearest base for 113 is 100.

$113 = 100 + 13$

Excess from the base = 13

1st part = Number + 2 × Excess

= 113 + 2 × 13

= 139

2nd part = New excess × Original excess

= (139 – 100) × 13

= 39 × 13

= 507

3rd part = Cube of excess

= 13^3

= 2197

Arrange the three parts as follows

139 | **5** 07 | **21** 97

Hence $(113)^3 = 1442897$

Example: Find the cube of 109 by Yavadunam Sutra

Solution:

The nearest base for 109 is 100.

109 = 100 + 09

Excess from the base =09

1st part = Number + 2 × Excess

= 109 + 2 × 9

= 127

2nd part = New excess × Original excess

= (127 – 100) × 9

= 27 x 9

= 243

3rd part = Cube of excess

= 9^3

= 729

Arrange the three parts as follows –

= **127** | **2** 43 | **7** 29

= 1295029

Exercise for Practice

Exercise 8.1

Find the Cube of the following by Anurupyena Vedic Method.

(a) 14	(b) 17	(c) 21
(d) 35	(e) 47	(f) 54
(g) 67	(h) 76	(i) 76
(j) 88	(k) 94	(l) 99

Exercise 8.2

Find the Cube of the following by NIkhilam Vedic Sutra.

(a) 13	(b) 19	(c) 25
(d) 36	(e) 46	(f) 54
(g) 69	(h) 87	(i) 104
(j) 113	(k) 208	(l) 315

Exercise 8.3

Find the Cube of the following by Ekadhikena Purvena Vedic Sutra.

(a) 29	(b) 34	(c) 44
(d) 55	(e) 63	(f) 78
(g) 86	(h) 97	(i) 98

Exercise 8.4

Find the Cube of the following by Yavadunam Vedic Method.

(a) 16	(b) 13	(c) 17
(d) 94	(e) 96	(f) 97
(g) 99	(h) 102	(i) 105

❑❑❑

9

CUBE ROOT

Introduction

The cube of a number is that number raised to the power 3. Thus, the cube of **a** is $\mathbf{a^3}$.

Suppose $a^3 = x$

$$a = (x)^{\frac{1}{3}}$$

There is no such method in our school curriculum which is useful enough to find the cube root of any number. Generally, we use the factor method to extract the cube root of any number, but that too is not good enough to find the cube root of any number having 7, 8 or more digits. Vedic Mathematics has its solution. There are around 4 or 5 Vedic methods written in the original work of Sri Swami Krishna Tirthaji Maharaj.

The great Indian astronomer and mathematician, Aryabhatta has also mentioned about a method to extract cube root of any number in his book **Ganita Pada** but the method is too complex to understand. The fifth sloka of **Aryabhatta's** book **Ganita Pada** reads as follows.

अघनाद भजेद द्वितीयात त्रिगुणेन घनस्य मूलवर्गेण ।
वर्ग स्त्रिपूर्व गुणितः शोध्यः प्रथमाद धनस्य घनात् ॥

Before Aryabhatta, there was no proper method to extract the cube root of any number. All the mathematicians after him followed his method with some modifications here and there. This elegant method described by

Aryabhatta has much symmetry with the Vedic method used by Swami Bharati Krishna Tirthaji Maharaj. Henceforth, I shall focus on only the Vedic method and not on Aryabhatta's method. The detailed meaning of the sloka is given here.

Let us call the unit place as Ghana, the 10-place as **the first Aghana** and the 100- place as the **Second Aghana** and so on. Let us see the following table.

Second Aghana	First Aghana	Unit
10^2	10	1
10^4	10^3	
____	____	
____	____	

The steps to find the cube root of a number on the basis of the sloka are as follows:

1. Subtract the greatest possible cube from the last Ghana place.
2. Divide the second Aghana place by thrice the square of the cube root already obtained in step 1.
3. Subtract from the first Aghana place the square of the quotient multiplied by thrice of the previous cube root.
4. Subtract the cube of the quotient from the Ghana place.
5. Repeat the process until all the digits are exhausted.

Important Points to Remember

- The cubes of first nine natural numbers are as follows:

M	1	2	3	4	5	6	7	8	9
M^3	1	8	27	64	125	216	343	512	729

- The number of digits in a cube root is the same as the number of 3-digit groups in the original cube including a single digit or a double-digit group if there exists any.
- 1, 4, 5, 6, 9 and 0 repeat themselves in the cube-ending.
- 2, 3, 7 and 8 have an inter-play of complements from 10.

In other words, the cube of 2, 3, 7 and 8 ends with 8, 7, 3 and 2 respectively.

Unit digit of a cube root	Unit digit of a cube
1	1
8	2
7	3
4	4
5	5
6	6
3	7
2	8
9	9

- The first digit (from left) of the cube root will always be obvious from the first group in the cube.
- Thus, the number of digits, the first digit and the last digit of the cube root of and exact cube are the data with which we start when we begin the work of extracting the cube root of an exact cube.
- Most importantly, the digit sum of any number (digit sum of a number can be obtained by using the casting out nines method described in the book in detail. please refer to it for more information) will always be 1, 8 or 9(0).

  Example: The digit sum of 314432 is 3 + 1 + 4 + 4 + 3 + 2 = 8

  Therefore 314432 is the perfect cube

Vedic method of finding the cube root of any number

We shall categorise the method into two parts:

1. ***Cube Root of a number having less than 7 digits***
2. ***Cube Root of a number having 7, 8 or 9 digits.***

Case 1: Finding the cube root of a number having less than 7 digits

This is done by inspection method. This method consists of the following steps:

- Look at the digit at the unit place and determine the unit place using the above table.
- Place a bar from right to left, leaving 2 digits at a time.

  For example

  $$\overline{3}\,43\,\overline{1}\,33\,\overline{1}$$

- If no digit is left (as in case of 343), the digit obtained in step 1, is the required cube root.
- If digits are left, find the largest single digit number whose cube root is less than or equal to this left over number.

Let us take a few examples.

Example : Find the cube root of 1728.

Solution: The extraction of the cube root of 1728 is shown in the following steps.

1. Place the bar over the number. $\overline{1}\,7\,2\,\overline{8}$
2. Since the bar is placed on two numbers, the cube root of this number contains two digits.
3. The first bar falls on unit digit 8. Looking upon the table, we can say that the unit digit of the cube root is 2.
4. The next bar falls on 1, and 1 is the largest number whose cube is less than 2. Therefore, the ten's digit of the cube root is 1.

Hence $\sqrt[3]{1728} = 12.$

Example: Find the cube root of 35937.

Solution:

1. Place the bar over the number from right to left, leaving two digits at a time.

   $$3\,\overline{5}\,9\,3\,\overline{7}$$

2. The bar falls on the unit digit 7, therefore the unit digit of the cube root is 3 (see Table for the reference)

3. The ten's digit of the cube root is 3, because

$3^3 < 35 < 4^3$

Hence, $\sqrt[3]{35937} = 33$

Example: Find the cube root of 389017

Solution:

1. Place the bar over the number from left to right, leaving two digits at a time.

$$38\overline{9}01\overline{7}$$

2. Since the bar is placed on two numbers, hence the cube root will contain only two digits.

3. Since the unit digit of this number is 7, on comparing the table shows that the unit digit of this number is 3.

4. For the ten's digit, take the least of

$7^3 < 389 < 8^3$

Hence the ten's digit = 7

$\sqrt[3]{389017} = 73$

Case 2: Cube root of a number having more than 7 digits but less than 10 digits

Rule:

1. A number with 7, 8 or 9 digits will have 3 digits in its cube root.
2. Denote the left digit by L, the middle digit by M and the right digit by R.
3. Subtract R^3 from the number and forget the last digit (0) obtained after subtraction.
4. The middle digit of the cube is obtained by $3\ R^2 M$.
5. The cube root of such number = L M R.

Example: Find the cube root of 33076161

Solution:

1. Placing the bar over the number we find that there are three groups of numbers, so the cube root will have three digits. Thus N = 3

$$\sqrt[3]{3\bar{3}07\bar{6}16\bar{1}}$$

2. From the table, the unit digit of the cube root is 1.

   Hence R = 1. More over $3^3 < 33 < 4^3$ 1, so L = 3

3. Subtract R^3 from the number and forget the last digit (0) obtained after subtraction.

$$\begin{array}{r} 3\,3\,0\,7\,6\,1\,6\,1 \\ -1 \\ \hline 3\,3\,0\,7\,6\,1\,6\,\not{0} \end{array}$$

4. Middle digit of the cube root is obtained by 3 R^2 M.

$$3\ R^2\ M = 3 \times 1^2 \times M = 3\ M$$

5. Now, suppose that this 3 R^2 M is contained in 3 3 0 7 6 1 6. Since 3 R^2 M = 3M is contained in 3 3 0 7 6 1 6, we may say that both should end with the same digit. Last digit of 3307616 is 6, hence, M = 2

6. We have thus got

   L = 3 M = 2 and R = 1

$$\sqrt[3]{3\bar{3}07\bar{6}16\bar{1}} = 321$$

Example: Find the cube root of 84604519

Solution:

1. Placing the bar over the number we find that there are three groups of numbers, so the cube root will have three digits. So N = 3

$$\sqrt[3]{8\bar{4}60\bar{4}51\bar{9}}$$

2. From the table, the unit digit of the cube root is 9.

   Hence R = 9. Moreover, $4^3 < 84 < 5^3$, so L = 4

3. Subtract R^3 from the number and forget the last digit (0) obtained after subtraction.

$$\begin{array}{r} 8\,4\,6\,0\,4\,5\,1\,9 \\ -\,7\,2\,9 \\ \hline 8\,4\,6\,0\,3\,7\,9\,\not{0} \end{array}$$

4. Middle digit of the cube root is obtained by $3\ R^2\ M$.

   $3\ R^2\ M = 3 \times 9^2 \times M = 243\ M$

5. Now, suppose that this $3\ R^2\ M$ is contained in 8 4 6 0 3 7 9. Since $3\ R^2\ M = 2\ 4\ 3\ M$ is contained in 8 4 6 0 3 7 9, we may say that both should end with the same digit. Last digit of 8 4 6 0 3 7 9 is 9, hence, M = 3

6. We have thus got

   L = 4 M = 3 and R = 9

$$\sqrt[3]{8\,\overline{4}\,6\,0\,\overline{4}\,5\,1\,\overline{9}} = 439$$

Example: Find the cube root of 143055667

Solution:

1. Placing the bar over the number we find that there are three groups of numbers, so the cube root will have three digits. Thus N = 3

$$\sqrt[3]{1\,4\,\overline{3}\,0\,5\,\overline{5}\,6\,6\,\overline{7}}$$

2. From the table, the unit digit of the cube root is 7.

   Hence R = 3. Moreover, $5^3 < 33 < 6^3$, so L = 5

3. Subtract R^3 from the number and forget the last digit (0) obtained after subtraction.

$$\begin{array}{r} 1\,4\,3\,0\,5\,5\,6\,6\,7 \\ -\,2\,7 \\ \hline 1\,4\,3\,0\,5\,5\,6\,4\,\not{0} \end{array}$$

4. Middle digit of the cube root is obtained by $3\ R^2\ M$.

$$3\ R^2\ M = 3 \times 3^2 \times M = 27\ M$$

5. Now, suppose that this $3\ R^2\ M$ is contained in 1 4 3 0 5 5 6 4. Since $3\ R^2\ M = 2\ 7\ M$ is contained in 1 4 3 0 5 5 6 4 , we may say that both should end with the same digit. Last digit of 1 4 3 0 5 5 6 4 is 4, hence, M = 2

6. We have thus got

$$L = 5 \qquad M = 2 \quad \text{and} \quad R = 3$$

$$\sqrt[3]{14305567} = 523$$

The above method is adopted mainly for odd cubes. If the cube be even, ambiguous values may arise at each step and tend to confuse the mind of the readers. Now the question is how to overcome this problem?

All you need to do is to go on dividing the given number by 8, until an odd cube emanates. Now apply the above method and multiply by the proper multiplier thereafter.

Example: Find the cube root of 48228544.

Solution: The number of digits of 48228544 is 8, which is an even; hence there is an ambiguity in this problem. Let us first understand the ambiguous case and then solve it with the new method discussed.

1. First group · 4 8 2 2 8 5 4 4.

$$4\overline{8}22\overline{8}54\overline{4}$$

2. Here, N = 3, L = 3 , R = 4 and M = ?

3. Subtract R^3 from the number and avoid the last digit.

$$\begin{array}{r} 4\,8\,2\,2\,8\,5\,4\,4 \\ -\,6\,4 \\ \hline 4\,8\,2\,2\,8\,4\,8\,\not{0} \end{array}$$

4. Middle digit of the cube root is obtained by $3R^2M = 3 \times 4^2 \times M = 48\ M$

5. The value of M in 48M should be such that, it contains 8 at the end. For M = 1 and M= 8, we get the last digit of the product = 8, now the ambiguous case arises.

How to avoid ambiguous case

As mentioned earlier, when the number whose cube root has to be extracted is even, divide it with 8, until an odd cube emanates.

$$\begin{array}{r|l} 8 & 48228544 \\ \hline 8 & 6028568 \\ \hline & 753571 \end{array}$$

Now find the cube root of 753571.

1. Place the bar

$$75\overline{3}7\overline{1}$$

2. Unit digit of the cube root = 1, and the second digit of the cube root is 9 because

   $9^3 < 753 < 10^3$,

3. Hence cube root of 753571 is 91.

   $\sqrt[3]{48228544} = \sqrt[3]{8 \times 8 \times 753571}$

   $= 2 \times 2 \times 91 = 364.$

Example: Find the cube root of 1906624

Solution: Since this is an even number, hence to avoid the ambiguous case in extracting the cube root, we need to divide the number by 8 until an odd cube emanates.

$$\begin{array}{r|l} 8 & 1906624 \\ \hline 8 & 238328 \\ \hline & 29791 \end{array}$$

The cube root of 29791 can be extracted merely by inspection as done in case 1, because it contains less than 7 digits.

Group the number

$$2\overline{9}79\overline{1}$$

Since there are two groups, so the cube root of 29791 contains a two-digit number.

Unit digit = 1 and ten's digit = 3

Hence, the cube root of 29791 = 31

We have,

$\sqrt[3]{1906624} = \sqrt[3]{8 \times 8 \times 29791}$

$= 2 \times 2 \times 31 = 124$

Exercise for Practice

Exercise 9

Find the Cube Root of the following numbers.

(a) 6892	(b) 636056
(c) 314432	(d) 8365427
(e) 1061208	(f) 8489664
(g) 143055667	(h) 9800344
(i) 8615125	(j) 33076161
(k) 2196	(l) 83453453
(m) 143877824	(n) 830584
(o) 300763	(p) 970299

□□□

10

SIMPLE EQUATIONS

Introduction

The algebraic method of solving an equation is itself a horrific experience for the common student. The amount of calculations done to reach the answer is a labyrinthine process. The Vedic sutra helps us immensely to reach the answer in a moment and that too without using pen and paper.

Don't you think to get rid of time-consuming methods being taught in modern-day schools?

Your problem is a reality which everyone faces, there being no such fool-proof method which can help you out from the dragon's mouth. But don't worry; here we shall deal with some special type of algebraic equation which can be solved merely by inspection.

Vedic Sutras for solving Simple Equation

Though there are a number of Vedic Sutras which practically deal with different types of equations, here we shall deal with only three Vedic Sutras and their different implications under different situations. These Vedic Sutras will help you to solve the Simple equation, Quadratic equation and Simultaneous equation merely by inspection. Once in running practice, this will prove as panacea for solving the Algebraic Equation.

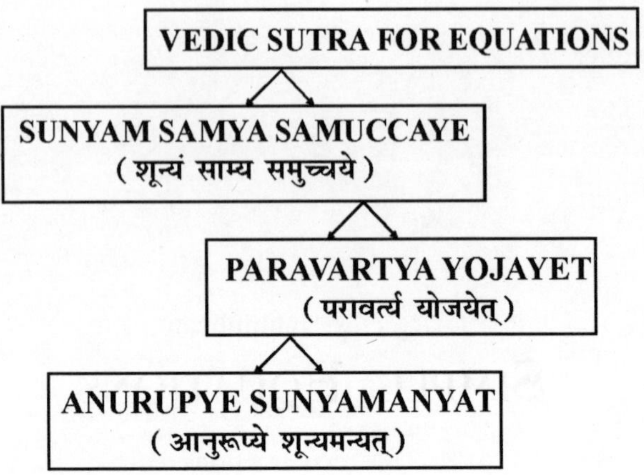

Meaning of Vedic Sutra

1. **Sunyam Samya Samuccaye**(शून्यं साम्य समुच्चये): -This sutra is applicable to a large number of different cases. It literally means - when the samuccaya is the same, equal it to zero. Samuccaya is a technical term which has several meanings under different contexts; and we shall explain them one by one.
2. **Paravartya Yojayet**: (परावर्त्य योजयेत्): - The literal meaning of this Vedic subtraction-sutra is – **transpose and adjust**. As far as the application of this sutra is concerned, it has numerous applications and is splendidly useful.
3. **Anurupye Sunyamanyat** (आनुरूप्ये शून्यमन्यत्): - This sutra is useful in solving a special type of simultaneous equation when the ratio of coefficients of either x or y is equal to the ratio of the constant terms. The literal meaning is – **If one is in ratio, the other one is zero**. In other words, if the ratio of x is equal to the ratio of constant terms, then the value of y will be zero. Similarly, if the ratio of y is equal to the ratio of constant term, then the value of x will be zero.

Application of Sunyam Samya Samuccaye (शून्यं साम्य समुच्चये)

Let us take this example and solve it, first traditionally and then by means of Vedic sutra and see the difference.

CASE 1

Example:

$$\frac{1}{x-3}+\frac{1}{x-5}=0$$

Solution:

$$\frac{x-5+x-3}{(x-3)(x-5)}=0$$

$\Rightarrow$ $2x - 8 = 0$ (By cross multiplication)

$\Rightarrow$ $2x = 8$

$\Rightarrow$ $x = 4$

Vedic Method: When the two fractions having the same numerical numerator is to be solved: the Samuccaya sutra says to equate the sum of denominator to zero.

The numerator having the same value, the Sunyam Sam Samuccaya sutra applies here:

Sum of Denominator = $x - 3 + x - 5 = 0$

$\Rightarrow$ $2x - 8 = 0$

$\Rightarrow$ $x = 4$

Example: $\mathbf{\frac{1}{2x-1}+\frac{1}{3x-4}=0}$

Solution: Since the numerator of the above fraction is numerically equal, hence after applying the Vedic sutra we get,

Sum of denominator = $2x - 1 + 3x - 1 = 0$

$\Rightarrow$ $5x - 2 = 0$

$\Rightarrow$ $x = 2 /5$

Example: Solve $\mathbf{\frac{6}{x+2}+\frac{6}{x+3}=0}$

Solution: The numerator of the given equation is numerically equal hence the Vedic sutra Sunyam Sam Samuccaya applies here too.

Sum of Denominator = $2x + 5 = 0$

Or, $x = -5 / 2$

CASE 2

Example: Solve $\frac{2x+9}{2x+7} = \frac{2x+7}{2x+9}$

Solution: The algebraic solution is left for the readers. The second meaning of the Vedic Sutra is – when the sum of numerators and denominators are same, then that sum should be equal to zero.

Here,

$N_1 + N_2$ = Sum of Numerators = 2x + 9 + 2x + 7 = 4x + 16

$D_1 + D_2$ = Sum of Denominator = 2x + 7 + 2x + 9 = 4x + 16

Since $N_1 + N_2 = D_1 + D_2$

This Vedic Sutra is applicable here too.

$N_1 + N_2 = D_1 + D_2 = 0$

$\Rightarrow$ 4x + 1 6 = 0

$\Rightarrow$ x = – 4

Example: Solve $\frac{3x+4}{3x+1} = \frac{x+1}{x+4}$

Solution:

$N_1 + N_2$ = Sum of Numerators = 3x + 4 + x + 1 = 4x + 5

$D_1 + D_2$ = Sum of Denominator = 3x + 1 + x + 4 = 4x + 5

Since $N_1 + N_2 = D_1 + D_2$

This Vedic Sutra is applicable here too.

$N_1 + N_2 = D_1 + D_2 = 0$

$\Rightarrow$ 4x + 5 = 0

$\Rightarrow$ x = – 5/4

Example: Solve $\frac{3x+2}{5x+7} = \frac{x+1}{3x-1}$

Solution: If in the algebraic total, there be a numerical factor, that should be removed.

Here,

$N_1 + N_2$ = Sum of Numerators = $3x + 2 + x + 1 = 4x + 3$

$D_1 + D_2$ = Sum of Denominator = $5x + 7 + 3x - 1 = 8x + 6 = 2(4x + 3)$

Removing the numerical factor from the sum of denominator we find

$$N_1 + N_2 = D_1 + D_2$$

Hence, this Vedic Sutra is also applicable here too.

$$N_1 + N_2 = D_1 + D_2 = 0$$

$\Rightarrow$ $4x + 3 = 0$

$\Rightarrow$ $x = -3/4$

CASE 3

So far, in the two cases we have dealt with, it will be observed that cross- multiplication of the coefficients of x gives us the same coefficients for x^2. There is the third case, when the cross multiplication of LHS and RHS gives the different coefficient of x^2, then in general term we may refer such type of equation as Quadratic Equation. It is well-known that a quadratic equation has two roots. In such case

$N_1 + N_2 = D_1 + D_2 = 0$, gives the first root of equation

And, $N_1 - D_1 = N_2 - D_2 = 0$, gives the second root of equation

Example: Solve $\dfrac{3x+4}{6x+7} = \dfrac{5x+6}{2x+3}$

Solution: On cross multiplication, we find that the coefficient of x^2 in LHS and RHS are respectively 6 and 30, which are different. By applying the third case of Sunya Samya Samuccaya, we get the two roots of this quadratic equation.

For the first root

$$N_1 + N_2 = D_1 + D_2 = 0$$

$N_1 + N_2 = 3x + 4 + 5x + 6 = 8x + 10$

$D_1 + D_2 = 6x + 7 + 2x + 3 = 8x + 10$

$\Rightarrow$ $8x + 10 = 0$

or, $x = -5/4$

For the second root

$N_1 - D_1 = N_2 - D_2 = 0$

$N_1 - D_1 = 3x + 4 - 6x - 7 = 0$

$\Rightarrow \quad -3x - 3 = 0$

$x = -1$

or, $N_2 - D_2 = 5x + 6 - 2x - 3 = 0$

$3x + 3 = 0$

$x = -1$

Hence the two roots of the above equation are x = – 5/4 and – 1.

Example: Solve $\dfrac{3x+6}{6x+3} = \dfrac{5x+4}{2x+7}$

Solution: On simple observation, it is evident that the above equation is a quadratic equation. We now apply the third meaning of Samuccaya sutra to obtain the two roots of the quadratic equation.

For the first root

$$N_1 + N_2 = D_1 + D_2 = 0$$

$N_1 + N_2 = 3x + 6 + 5x + 4 = 8x + 10$

$D_1 + D_2 = 6x + 3 + 2x + 7 = 8x + 10$

$\Rightarrow \quad 8x + 10 = 0$

or, $x = -5/4$

For the second root

$\Rightarrow \quad N_1 - D_1 = N_2 - D_2 = 0$

$\Rightarrow \quad N_1 - D_1 = 3x + 6 - 6x - 3 = 0$

$\Rightarrow \quad -3x + 3 = 0$

$\Rightarrow \quad x = 1$

or, $N_2 - D_2 = 5x + 4 - 2x - 7 = 0$

$\Rightarrow \quad 3x - 3 = 0$

$\Rightarrow \quad x = 1$

Hence the two roots of the above equation are x = – 5/4 and 1.

CASE 4

When the sum of numerator and denominator in LHS and RHS are same, equate the sum total of denominator to zero.

Example: $\frac{1}{x-8}+\frac{1}{x-9}=\frac{1}{x-12}+\frac{1}{x-5}$

Solution:

$N_1 + N_2$ of LHS = $N_1 + N_2$ of RHS = 1 + 1 = 2

$D_1 + D_2$ of LHS = x − 8 + x − 9 = 2x − 17

$D_1 + D_2$ of RHS = x − 12 + x − 5 = 2x − 17

By Samuccaya Vedic Sutra

$D_1 + D_2 = 0$

2x − 17 = 0

x = 17 / 2

Example: $\frac{2}{x+2}+\frac{3}{x+3}=\frac{4}{x+4}+\frac{1}{x+1}$

Solution:

$N_1 + N_2$ of LHS = $N_1 + N_2$ of RHS = 2 + 3 = 5

$D_1 + D_2$ of LHS = x + 2 + x + 3 = 2x + 5

$D_1 + D_2$ of RHS = x + 4 + x + 1 = 2x + 5

By Samuccaya Vedic Sutra,

$D_1 + D_2 = 0$

2x + 5 = 0

x = - 5 / 2

Application of Paravartya Yojayet (परावर्त्य योजयेत्)

Though the Vedic sutra Paravartya Yojayet has different meanings and number of applications in different context, here we shall deal with only two of its applications which also have algebraic implication.

CASE 1: For the given equation

$$(x + a)(x + b) = (x + c)(x + d)$$

$$x = \frac{cd - ab}{a + b - c - d}$$

Example: $(x + 1)(x+2) = (x - 3)(x - 4)$

Solution: Applying the above sutra,

$$x = \frac{12 - 2}{1 + 2 + 3 + 4} = 1$$

Example: $(x + 7)(x+9) = (x + 3)(x + 22)$

Solution: Applying the above sutra,

$$x = \frac{66 - 63}{7 + 9 - 3 - 22} = \frac{-3}{9} = \frac{-1}{3}$$

Example: $(x - 2)(x - 5) = (x -1)(x -4)$

Solution: Applying the above sutra,

$$x = \frac{4 - 10}{-2 - 5 + 1 + 4} = \frac{-6}{-2} = 3$$

CASE 2:

For the given equation

$$\frac{m}{x + a} + \frac{n}{x + b} = 0$$

$$x = -\left[\frac{mb + na}{m + n}\right]$$

Example: Solve, $\frac{4}{x + 2} + \frac{3}{x + 5} = 0$

Solution:

$$x = \frac{-(20 + 6)}{4 + 3}$$

$$x = \frac{-26}{7}$$

Example: Solve, $\frac{7}{x-3}+\frac{5}{x+5}=0$

Solution:

$$x=\frac{-(35-15)}{7+5}$$

$$x=\frac{-20}{12}=\frac{-5}{3}$$

Example: Solve, $\frac{11}{x+8}+\frac{2}{x-5}=0$

Solution:

$$x=\frac{-(-55+16)}{11+2}$$

$$x=\frac{39}{13}=3$$

CASE 3 : Solving Simultaneous Equations

Paravartya Sutra is also applicable in solving linear equations of two variables. The methods taught in our school to solve linear equations are either lengthy or confusing. Though the Cross-Multiplication method for solving the linear equation of two variables is the best method of the current system but the major drawback of this method is regarding the plus and minus sign. Students and even teachers sometimes get confused over the plus or minus sign. The *Paravartya* rule enables us to give the correct answer mentally without any confusion.

Consider the system of linear equation in two variables.

$$a_1x + b_1y + c_1 = 0$$

$$a_2x + b_2y + c_2 = 0$$

The value of × and y can be extracted using Cross-Multiplication Method as follows which is nothing but the extension of Paravartya Sutra.

$$\frac{x}{\begin{vmatrix} b_1 & c_1 \\ b_2 & c_2 \end{vmatrix}}=\frac{y}{\begin{vmatrix} c_1 & a_1 \\ c_2 & a_2 \end{vmatrix}}=\frac{1}{\begin{vmatrix} a_1 & b_1 \\ a_2 & b_2 \end{vmatrix}}$$

This is usually written as

$$x = \frac{b_1c_2 - b_2c_1}{a_1b_2 - a_2b_1} \quad \text{and} \quad y = \frac{c_1a_2 - c_2a_1}{a_2b_2 - a_2b_1}$$

3. Application of Anurupyena Sunyam Anyat (आनुरूप्ये शुन्यमन्यत्)

This is a special case of solving linear equation of two variables. Let me take an example which will make the meaning of this sutra clear.

Example: Solve the following equation for x and y

$12x + 8y = 7$

$16x + 16y = 14$

Solution:

Here the ratio of y is 8 : 16 = 1 : 2 and the ratio of constant term is 7 : 14 = 1 : 2.

Since the ratio of y is equal to the ratio of constant term hence the value of x =0, because the sutra says –

If the ratio of coefficients of either x or y is equal to the ratio of the constant term, then the value of the other term i.e. y or x is zero. In other words, if the ratio of coefficients of x is equal to the ratio of the constant term, then the value of y is zero and vice-versa.

Here the ratio of y is 1:2 and the ratio of the constant term is 1: 2 hence the other term x = 0, put the value of x in any equation and get the value of y instantly.

For x = 0, y = 8/7

Example: Solve for x and y

$4x + 6y = 12$

$12x + 8y = 36$

Solution:

The ratio of coefficients of x = 4 : 12 = 1 : 3

The ratio of coefficients of y = 6 : 8 = 3 : 4

The ratio of constant term = 12 : 36 = 1 : 3

Here the ratio of coefficient of x and the ratio of constant term are equal so the value of y = 0.

Put the value of y in any equation to get x = 3.

Example: Solve for x and y

499x + 172y = 212

9779x + 387y = 477

Solution:

The ratio of coefficients of x = 499 : 9779

The ratio of coefficients of y = 172 : 387 = 4 : 9

The ratio of constant term = 212 : 477 = 4 : 9

Here the ratio of coefficient of y and the ratio of constant term are equal so the value of x = 0.

Put the value of x in any equation to get y = 53/43.

Example:: Solve for x and y

99x + 72y = 396

79x + 144y = 316

Solution:

The ratio of coefficients of x = 99 : 79

The ratio of coefficients of y = 72 : 144 = 1:2

The ratio of constant term = 396 : 316 = 99 : 79

Here the ratio of coefficient of x and the ratio of constant term are equal so the value of y = 0.

Put the value of y in any equation to get x = 4.

Exercise for Practice

Exercise 10

Solve the following Equation using appropriate Vedic Sutra

(a) $\frac{1}{x+4}+\frac{1}{x-6}=0$

(b) $\frac{5}{3x+2}+\frac{5}{2x+8}=0$

(c) $\frac{2x+4}{2x+1}=\frac{2x+1}{2x+4}$

(d) $\frac{3x+2}{5x-7}=\frac{x+1}{3x+13}$

(e) $\frac{5x+7}{2x+3}=\frac{x+1}{4x+5}$

(f) $\frac{3x+6}{2x+5}=\frac{5x+4}{6x+5}$

(g) $\frac{1}{x+2}+\frac{1}{x+6}=\frac{1}{x+1}+\frac{1}{x+7}$

(h) $12x + 42y = 7$
$16x + 96y = 16$

(i) $5x + 11y = 15$
$10x + 44y = 30$

❑❑❑

ANSWER SHEET

Ch–1 Addition

Ex 1.1

1. 212790
2. 2122736970
3. 1558199
4. 12552650
5. 34079177437664

Ex 1.2

1. 25141812
2. 31401308261
3. 2790
4. 39899
5. 23189875

Ex. 1.3

1. 838.568 Km
2. 105.641 Kg
3. Rs. 1306.90
4. Rs. 1892.51

Ch – 2 Subtraction

Ex. 2.1

(1) 970712
(2) 407588655
(3) 41745909160653
(4) 1660307
(5) 2331
(6) 10092
(7) 196
(8) 288

Ex. 2.2

1) 18288258
2) 35938897
3) 108233
4) 2287895213
5) 191072
6) 6477871
7) 3443
8) 27429

Ex. 2.3

1. 184	2. 1182	3. 297000567
4. 45675782	5. 115087	6. 457574
7. 376899826	8. 109091	

Ch – 3 Multiplication

Ex. 3.1

a) 270	b) 4048	c) 13860
d) 976128	e) 6693	f) 9792
g) 9310	h) 688344	i) 2425
j) 114540	k) 883452	l) 1019904
m) 1046902	n) 99910020	o) 99870012

Ex. 3.2

a) 2156	b) 478130	c) 3186
d) 74930	e) 478839	f) 483678

Ex. 3.3

a) 853146	b) 651348	c) 36466353
d) 548975451024	e) 6529745734702542	f) 45899541
g) 56488999943511	h) 64999935	i) 2543454
j) 55836	k) 45643311	l) 65399346

Ex. 3.4

a) 3456	b) 567018	c) 6174
d) 35290332	e) 4911769275	f) 368676
g) 484071	h) 35623341	i) 592184
j) 24007698	k) 20213248	l) 271466
m) 3441417512	n) 305368	

Ex. 3.5

a) 365056	b) 3807909	c) 1524
d) 211640	e) 2587344	f) 5701996

Ch – 4 Division

Ex. 4.1

a) Q = 5 R = 262	b) Q = 2390 R = 722
c) Q = 1 R = 3457	d) Q = 1R = 2204
e) Q = 233 R = 3711	f) Q = 22 R = 22068

Ex. 4.2

a) Q = 124 R = 111	b) Q = 10 R = 9
c) Q = 21 R = 4216	d) Q = 123 R = 6
e) Q = 22 R = 135	f) Q = 27 R = 98

Ex – 4.3

a) Q = 24	R = 24	b) Q = 102	R = 23
c) Q = 1356	R = 60	d) Q = 611	R = 97
e) Q = 73	R = 416	f) Q = 1323	R = 283
g) Q = 21555	R = 344	h) Q = 13219	R = 1866

Ch – 5 Casting Out Nines

a) Correct	b) Correct	c) Incorrect
d) Incorrect	e) Correct	f) Correct
g) Correct	h) Correct	i) Incorrect
j) Correct	k) Correct	l) Incorrect
m) Correct	n) Correct	o) Correct
p) Correct	q) Incorrect	r) Incorrect

Ch– 6 Square

Ex. 6.1

a) 225	b) 625	c) 1225
d) 4225	e) 7225	f) 9025
g) 11025	h) 13225	i) 15625

Ex. 6.2

a) 1936	b) 2209	c) 4624
d) 3249	e) 7921	f) 5776
g) 2809	h) 3481	i) 4481

Ex. 6.3

a) 225	b) 324	c) 1225
d) 2304	e) 2916	f) 3364
g) 8464	h) 12544	i) 11449

Ex. 6.4

a) 841	b) 1156	c) 2209
d) 1444	e) 3364	f) 6241
g) 7225	h) 9409	i) 9801
j) 11881	k) 28224	l) 39204
m) 5550736	n) 39112516	o) 67138281

Ch – 7 Square Root

Ex. 7.1

a) 2115	b) 3214	c) 3748
d) 46	e) 65	f) 83
g) 243	h) 354	i) 412
j) 1125		

Ex. 7.2

a) 9956	b) 965	c) 213
d) 889	e) 8887	f) 6732
g) 57	h) 743	i) 8361
j) 6423		

Ex. 7.3

a) 608	b) 12.8	c) 13.23
d) 6.57	e) 15.7	f) 14.07
g) 8.06	h) 16.94	i) 25.12

Ch – 8 Cube

Ex. 8.1

a) 2744	b) 4913	c) 9261
d) 42875	e) 103823	f) 157464
g) 300763	h) 438976	i) 681472
j) 830584		

Ex. 8.2

a) 2197	b) 6859	c) 15625
d) 46656	e) 97336	f) 157464
g) 328509	h) 658503	i) 1124864
j) 1442897	k) 8998912	l) 31255875

Ex. 8.3

a) 24389	b) 39304	c) 85184
d) 166375	e) 250047	f) 474552
g) 636056	h) 912673	i) 941192

Ex. 8.4

a) 4096	b) 2197	c) 4913
d) 830584	e) 884736	f) 912673
g) 970299	h) 1061208	i) 1157625

Ch – 9 Cube Root

a) 18	b) 86	c) 68
d) 203	e) 102	f) 204
g) 523	h) 214	i) 205
j) 321	k) 21	l) 437
m) 524	n) 94	o) 67
p) 99		

Ch – 10 Simple Equations

a) $x = 1$	b) $x = -2$	c) $x = -5/4$
d) $x = -3/4$	e) $x = -4/3$	f) $x = -5/4, -1$
g) $x = -4$	h) $x = 0, y = 1/6$	i) $x = 3, y = 0$

❑❑❑

BIBLIOGRAPHY

1. Vedic Mathematics – Swami Bharti Krisna Tirtha Ji
   Motilal Banarsi Das Publishers 1992
2. Indian Mathematics and Astronomy- S.Balachandra Rao
   Jnana Deep Publication Bangalore 1996
3. Topic on Vedic Mathematics taken from Mathematics
   Course book (Class IX and X) – Rajasthan Board
4. Dictionary of Mathematics- James and James
5. Ancient Indian Astronomy and Mathematics

The Joy of Mathematics

-- Sumita Bose

Filled with interesting games, activities, etc., built around real life experiences and based on mathematical concepts, this book it will appeal to both children as well as adults. There are amazing mathematical magic, short cut methods of mathematics, number patterns, optical illusions, puzzles and jokes.

Salient Features--

- ❖ Simple language
- ❖ Picture illustrations for easy understanding
- ❖ Practical application of mathematics.

Parents and teachers can use this book to motivate their children/students to enjoy mathematics.

UNICORN BOOKS

F-2/16, Ansari Road, Daryaganj, New Delhi-110002

☎ 23275434, 23272783/84 • Fax: 011&23257790

E-mail: unicornbooks@vsnl.com • Website: www.unicornbooks.in

गणित में सरसता और आनन्द

–लायकराम शर्मा

वर्तमान कम्प्यूटर युग में 'गणित' का व्यावहारिक महत्व बहुत बढ़ गया है। आज गणित में प्रगति की संभावनाएं असीमित हैं। इसी क्रम में प्रस्तुत पुस्तक 'गणित में सरसता और आनन्द' उच्च एवं उच्चतर कक्षाओं के विद्यार्थियों के साथ-साथ जन-साद्यारण को भी मनोरंजन के माध्यम से 'गणित' विषय की ओर आकर्षित करेगी। जनसाधारण की भाषा में लिखी इस पुस्तक में –

- ❖ मानसिक व्यायाम (Mental Exercises) हेतु रोचक प्रश्नों का वर्णन है।
- ❖ मनोरंजक गणितीय क्रम (Mathematical Patterns) का उल्लेख है।
- ❖ वर्ग और घन निकालने की सूक्ष्म विधि (Shortcuts) है।
- ❖ भारत के महान गणितज्ञों पर विवेचना।

आशा ही नहीं बल्कि पूर्ण विश्वास है कि यह पुस्तक शिक्षकों एवं विद्यार्थियों के लिए उपयोगी साबित होगी।

F-2/16, अंसारी रोड, दरियागंज, नई दिल्ली-110002

☎ 23275434, 23272783/84 • फैक्स: 011-23257790

E-mail: unicornbooks@vsnl.com • *Website:* www.unicornbooks.in